From Publishing to Knowledge Networks

Springer
Berlin
Heidelberg
New York
Hong Kong
London
Milan
Paris
Tokyo

Alexander Hars

From Publishing to Knowledge Networks

Reinventing Online Knowledge Infrastructures

With 29 Figures and 31 Tables

Springer

Professor Dr. Alexander Hars
University Bayreuth
Lehrstuhl BWL 7
Gebäude B 9
95440 Bayreuth
Germany

ISBN 3-540-01250-8 Springer-Verlag Berlin Heidelberg New York

Cataloging-in-Publication Data applied for
A catalog record for this book is available from the Library of Congress.
Bibliographic information published by Die Deutsche Bibliothek
Die Deutsche Bibliothek lists this publication in the Deutsche Nationalbibliografie; detailed bibliographic data is available in the Internet at <http://dnb.ddb.de>.

Springer-Verlag Berlin Heidelberg New York
a member of BertelsmannSpringer Science+Business Media GmbH

http://www.springer.de

Printed in Germany

Hardcover-Design: Erich Kirchner, Heidelberg

SPIN 10922828 42/3130-5 4 3 2 1 0 – Printed on acid-free paper

Preface

Information technology will have a significant impact in the publishing industry. Many publishers and institutions are experimenting with new ways of sharing knowledge. Internet sites are becoming more sophisticated, digital libraries and collaboratories are being built. However it is still unclear, where this journey of change will take us. This book sets out to provide answers by examining a large number of different initiatives that aim to create, share and distribute knowledge electronically. The book also fundamentally rethinks the related processes and shows that information technology has the potential of being much more 'revolutionary' than just accelerating established publishing and review processes. The book shows that deeply ingrained notions of books and articles need to be rethought. In the future they may be replaced by dynamic networks of knowledge consisting of much smaller units than we are used to. The goal of this book is to provide guidance to those who are developing their own knowledge infrastructures and knowledge networks.

The book was written while working at the University of Southern California and I am indebted to my colleagues in the Information and Operations Management Department of the Marshall School of Business. The first ideas for this topic began to take form during a PhD seminar I taught together with Omar El Sawy. I thoroughly enjoyed the creative environment and lively discussions with him. I am grateful to Ann Majchrzak for opening many doors as well as for her critical thinking and systematic research approach from which I learned much. I also thank my German advisor, August-Wilhelm Scheer, for his encouragement and support. Thanks also to Il Im who accepted the burden of proofreading an early version as well as to my parents Christa and Owe. As always, writing took a toll on the family and I thank my wife Ulrike and my daughter Sabrina for their enduring patience and support.

Alexander Hars, April 2003

Content overview

Contents

List of abbreviations

ACM	Association of Computing Machinery
AI	Artificial intelligence
CRT	Cathode ray tube
CycL	CYC representation language
DNA	Deoxyribonucleic Acid
eEPC	Extended event-oriented process chain
EIL	Enterprise Engineering Lab (Toronto)
EMS	Electronic meeting system
FRISCO	Framework of information systems concepts
HTML	Hypertext markup language
IBM	International Business Machine Corporation
IFIP	International Federation of Information Processing
IEEE	Institute of Electrical and Electronics Engineers
IP	Internet protocol
IT	Information technology
IS	Information system
ISD	Information systems development
KIF	Knowledge Interchange Format
LAS	Los Alamos Sequence Library
JAIR	Journal of Artificial Intelligence Research
MIME	Multipurpose Internet Mail Extensions
MISQ	Management Information Systems Quarterly
MIT	Massachusetts Institute of Technology
PDF	Portable document format
SGML	Standard Generalized Markup Language
TOVE	Toronto Virtual Enterprise Project
URL	Uniform Resource Locator
WWW	World-wide web
XML	Extended Markup Language

Index of tables

Index of figures

1. Leveraging information technology for science

1.1. Motivation

Management theory increasingly recognizes the importance of knowledge as an additional production factor besides the classical factors (Gutenberg 1965) of labor, equipment and materials (Grant 1996; Kogut & Zander 1992; Nelson 1991; Drucker 1998). In the transition towards a knowledge based economy, information technology plays a crucial role. Thus many firms and institutions are currently reengineering their knowledge-intensive processes and are building IT-based knowledge infrastructures. Functional organization structures are being replaced by process-oriented structures. Processes are not only being shortened by reducing transfer times but also by parallelizing previously sequential tasks. Access to information and decision-making authority is being decentralized (while establishing centralized monitoring and exception handling mechanisms). In many cases, the concept of 'product' has been refocused around knowledge-intensive services. Moreover, many efforts are targeting the collection and organization of knowledge. Consulting companies, for example, are pioneering knowledge bases which organize and replicate knowledge about best practices (for an overview see Prusak 1998).

One of the areas where novel approaches for the management of knowledge have not yet taken a deeper hold are the sciences. Although they are very knowledge-intensive, their decentralized organization and the high complexity of scientific knowledge have been obstacles in the move towards more efficient management of scientific knowledge.

Many voices within the scientific community have recognized these problems and argued that better mechanisms for the accumulation and utilization of scientific knowledge are needed:

> “There is a growing mountain of research. But there is increased evidence that we are being bogged down today as specialization extends. The investigator is staggered by the findings and conclusions of thousands of other workers -- conclusions which he cannot find time to grasp, much less to remember, as they appear. Yet specialization becomes increasingly necessary for progress, and the effort to bridge between disciplines is correspondingly superficial. Professionally our methods of transmitting and reviewing the results of research are generations old and by now are totally inadequate for their purpose." ..."The difficulty seems to be, not so much that we publish unduly in view of the extent and

variety of present day interests, but rather that publication has been extended far beyond our present ability to make real use of the record." (Bush 1945)

Information technology provides many opportunities for addressing these problems. Many online scientific journals have been established which facilitate access to scientific knowledge, reduce publishing cost and shorten processing times for review and dissemination. However, as Hammer and Champy (1994) have argued, information technology must lead to more fundamental change than automating and accelerating traditional processes. Thus the capabilities of information technology may fundamentally change the scientific publishing system and the way in which scientists create, organize and disseminate knowledge. Such change is already visible in some online infrastructures, most notably in physics (Ginsparg 1994) and molecular biology (Benson et al. 1998), which differ in significant aspects from the traditional 'journal' model.

Nevertheless, most academic sites replicate traditional approaches which scientists have become accustomed to over the centuries. For example, articles continue to be treated as being immutable – although electronic articles could be updated easily. Many electronic journals still publish issues, although bundling electronic articles does not provide similar distribution advantages as bundling paper. Adhering to established structures is risky: reputable journals risk to be marginalized by nimble competitors that establish more effective sites. Moreover, better approaches to structuring and disseminating scientific knowledge may fundamentally change the way in which research is conducted and thus make research more productive and satisfying.

As the next two sections will show, the impact of new information technologies on the practice of science can hardly be overestimated. The next section discusses the close relationship between significant changes in information technology and scientific progress. To substantiate the argument, a brief analysis of potentials for change follows. This paves the way to refine the concept of sites which are dedicated to scientific knowledge ('scientific knowledge infrastructures') and to provide an overview over the subsequent chapters.

1.1.1. Impact of previous IT revolutions on science

The history of knowledge is inextricably linked with advances in information technology. As Table 1 shows, the first revolution in information technology was the invention of writing in Mesopotamia around 3500 B.C. (Jackson 1981, p 16; Van Doren 1991 p 10). It was an important factor for the emergence of the first sophisticated civilization. Writing allowed record-keeping and communication of rules and knowledge. Reading and writing led to new educational institutions. The creation and copying of manuscripts – usually in soft clay, wood, bamboo, papyrus or parchment – were costly. Thus the second revolution in information technology occurred when Ts'ai Lun invented paper at the turn of the first century. It was one of the key factors for the rapid rise of the Chinese civilization during the next millennium and only became known much later in the West.

The third revolution occurred around 1440 when Gutenberg invented the printing press. This new information technology dramatically reduced the costs of copying information and was a key enabler for the rise of science (Eisenstein 1983, pp 185-252). Before Gutenberg's invention Europe had been in stagnation for more than a millennium. The knowledge of ancient Greek philosophers had been lost and just rediscovered. The number of universities was small (see Figure 1). Eminent thinkers occupied themselves with theology and philosophy. Although paper had already become the dominant medium for the storage of knowledge, copying of manuscripts had to be performed manually. This was costly and error-prone. Direct access to knowledge was scarce and it was difficult to widely and rapidly disseminate ideas.

Table 1. Revolutions in information technology

Year	Technology
ca. 3500 BC	Writing invented in Mesopotamia
ca. 100 AD	Paper invented by Ts'ai Lun
ca. 1440 AD	Printing press invented by Gutenberg
ca. 1945 AD	Electronic computer

The printing press revolutionized the creation and dissemination of knowledge. Forty years after Gutenberg's invention, printing presses had sprung up everywhere in Europe (Steinberg 1959, p 28). Now the renaissance of ancient thought could be completed: "Within fifty years, nearly every important Greek and Roman work had been printed and distributed all over the learned world" (Van Doren 1991, p.154). The number of university foundings rose sharply (see Figure 1).

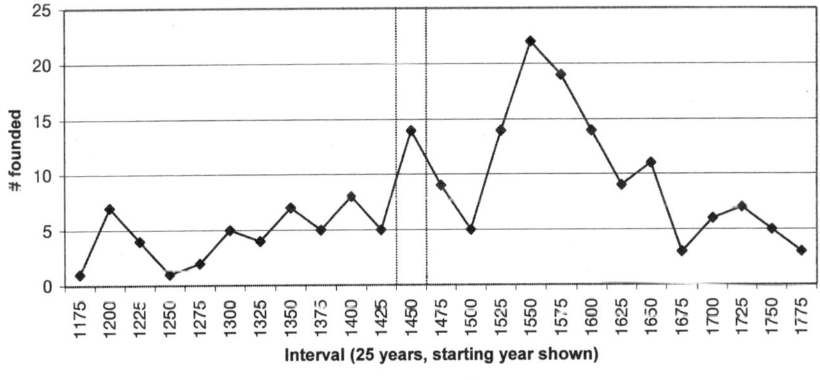

Figure 1. University foundings in Europe 1175-1799 (Data from: Frijhoff 1996, pp 90-94)

The amount of knowledge to which scholars had access exploded and the problems of correct translation and interpretation received much interest. Ideas could be published easily. In this climate, a new era of intellectual discourse began. The explosion of available knowledge enabled the development of new methods, processes and institutions for the creation of a reliable body of knowledge. It favored rational discourse and proof. As a consequence, modern science slowly emerged.

While information technology may not have been the only factor - movable type was known in China and Korea well before Gutenberg's invention (Hart 1992, p 43) - modern science would not have been possible without the printing press.

Currently, the next revolution in information technology is well under way. Like the previous revolutions it will lead to major changes. In contrast to prior revolutions, the causes are not only based on cost reduction, but also on novel capabilities. The introduction of paper reduced the costs of recording information by replacing more costly materials such as parchment, bamboo and papyrus. Similarly, the invention of the printing press reduced the costs and time needed for reproducing manuscripts. While computer technology further reduces the cost of storage and reproduction, it also provides entirely new capabilities: Information can be transmitted instantly over long distances; knowledge which is stored electronically can be modified even after the time of publishing. In addition, it is possible to interact with knowledge in the form of software and knowledge can be presented dynamically in different ways (e.g. as chart or figure) depending on the needs of the reader. Moreover, sound and moving images can be captured. It follows that the present revolution might lead to even more dramatic change than the previous two revolutions.

1.1.2. Potential changes to scientific knowledge

While information technology introduces many new capabilities, it is not clear to what degree scientific knowledge will be affected. A brief analysis may show some of the areas in which the characteristics of scientific knowledge may changes as a consequence of information technology.

Information technology may change the unit used for storing and manipulating scientific knowledge: In the paper environment, the main unit of academic knowledge is the article. An article is a collection of many thoughts and insights and typically contains between two thousand and eight thousand words. For economic reasons, paper journals cannot accept contributions that consist of just a few words or sentences. While it would be economically feasible to distribute such knowledge, the search and integration costs on the side of the reader are prohibitive. On the other hand, large book-sized contributions lead to significant publishing costs. As a consequence, paper journals are forced to limit the size of contributions and thereby may prevent relevant knowledge from being published. For example, evidence in the form of data is routinely suppressed from paper publications. In an electronic environment the costs of publication are nearly zero (see section 2.4.4.5.). This eliminates the upper boundary on article size. In addition, the ability to link information and provide alternative indexes and interfaces can substantially reduce the search costs for the reader. This reduces the limitations on the lower boundary of article size: the individual thought may become the smallest unit of publishable knowledge. In an online environment, therefore, the scientific body of knowledge may consist of a sea of individual scientific thoughts rather than a collection of articles.

In addition, information technology may change the form which scientific knowledge can take. Besides the addition of sound and moving images, scientific knowledge may become active in the form of software. Theories, models and artificial worlds can become alive on the screen. Early examples are already available: An online article by Dolin et al. (1988), for example, includes a section where readers can execute a text classification algorithm and apply it for their own purposes. Similarly a theory of financial decision-making might be formulated in software. The reader might no longer need to understand the theory in detail but could interact with it to apply it directly to his special case. Interactive scientific knowledge may change the dialogue between researcher and reader. It may eliminate the gap between theory and practice which has shielded research from application. At the same time, it may significantly raise the bar for the quality of a theory.

Furthermore, the electronic medium allows the utilization of algorithms for integrating knowledge into a larger body of knowledge and for checking the consistency between different contributions. Overlap with other contributions can be detected in many cases. It may be possible to generate questions aimed at improving structure and content of a contribution which at the same time can be used to check relationships with other contributions.

These changes to knowledge require the institutionalization of new approaches for the creation and dissemination of knowledge. As a consequence, the scientific community may need to adapt the processes of knowledge creation and dissemination to the new technology (Watson1994). At issue is not just information technology as a new medium for the storage of knowledge. Although the ability to modify stored knowledge sets it apart from traditional storage media, the bigger challenge is how to leverage the ability of information processing to improve the body of scientific knowledge. New methods and indexing systems must be accepted by the research community to be effective. While knowledge has traditionally been "pushed" at the audience, in the electronic environment readers can now pull together the knowledge they are interested in from various sites. The processing capability of computers can be used in many other ways to improve not only the retrieval but also the creation, integration and evaluation of knowledge.

This analysis has shown, that significant potential exists for reinventing the scientific communication system. In particular, online knowledge infrastructures may broaden their support from the dissemination of knowledge to the other phases of the research process, including creation, synthesis and evaluation of knowledge.

1.2. Analytical focus

This study is motivated by the transformation of the scientific publishing system which results from the emergence of computer technology. Unfortunately the term 'scientific publishing system' may not be well suited to describe the end product of this evolution. Information technology may change the focus away from publishing and dissemination towards discursive knowledge creation and synthesis. To fill this terminological void the term 'online scientific knowledge infrastructure' will be utilized in the following. The concept of 'infrastructure' is useful because it extends beyond technology. Technology forms the basis of an infrastructure but is never sufficient by itself. An infrastructure requires processes and links with the conventions of practice. It cannot be imposed but grows from the actions of many. An infrastructure must be adopted widely. For the individual, an infrastructure becomes part of the environment of which he becomes aware only when it fails (Star & Ruhleder 1996, p 113).

The current scientific publishing system has many characteristics of an infrastructure: it is widely accepted and most scientific activities are deeply embedded within it. For many researchers the publishing system has become so 'natural' that it is difficult to envision alternatives.

The object of this research thus is a future state of the scientific communication system where the potentials of information technology for the creation and dissemination of scientific knowledge are more fully exploited. While significant progress has been made, this state has not yet been reached. It requires both careful adaptation of information technology towards the needs of scientific processes and the fine-tuning of scientific processes towards the capabilities and requirements of information technology. Ideally, information technology would move into the background and assume the role of a substrate on which research and scientific communication can be performed. It would be invisible to the researcher who could exploit it in many different ways.

As a consequence, the term 'online scientific knowledge infrastructure' will be defined as:

> a socio-technical information system aimed at the creation, organization, storage and application of scientific knowledge. It is not only a technical system consisting of hardware, software, databases etc. It is also on organizational system which consists of stakeholders, organizational structures and processes.

The term 'knowledge infrastructure' needs to be distinguished from similar terms such as 'knowledge base' and 'knowledge repository'. Although 'knowledge base'/'knowledge repository' are occasionally used in the context of organization science to refer to the collective body of knowledge which an organization is able to bring to bear on its operations (Bierly & Chakrabarti 1996) they are predominantly used in a more technical sense. In product development, the term 'knowledge base' refers to computer systems which help engineers to reuse solutions for recurring problems (Borst, Akkermans & Top 1997). In AI it refers to a component of an expert system (or any system that imitates the reasoning behavior of humans). This component supplies 'knowledge' needed for solving a problem to a reasoning engine (Guida & Tasso 1994). A knowledge base or knowledge repository, thus, may be a component of a knowledge infrastructure.

Recently some authors have introduced the term 'knowledge network' (Director & King, 2001). This term can be regarded as synonym to knowledge infrastructure.

1.3. Objectives

Many scientific institutions have begun building online systems dedicated to scientific knowledge. They require a better understanding of the characteristics of online knowledge infrastructures and the choices which need to be made during their design. This research aims to provide a conceptual and empirical basis for the many efforts which are currently under way to use information technology to support the creation, integration, and dissemination of knowledge within science. The ultimate goal of this research is to increase the effectiveness of the scientific enterprise by providing a better understanding of the impact of information technology on the scientific communication system.

1.4. Approach

The second chapter examines characteristics of scientific knowledge infrastructures. This requires both a theoretical analysis of the role which information technology plays in scientific communication as well as an empirical analysis of current efforts which leverage information technology for improving the body of scientific knowledge. Therefore the chapter begins with a theoretical analysis of the transition from traditional paper-based journals towards online infrastructures. The impact of information technology on the main characteristics of paper-based journals and on the associated processes is evaluated systematically. The analysis will show that information technology has the potential of fundamentally changing the scientific communication system. In the second part of this chapter, an analysis of current online infrastructures follows. Internet sites which are dedicated to scientific knowledge are examined and their characteristics are identified. To complete the identification of characteristics, visions of future knowledge infrastructures that are different from the results of theoretical analysis and from currently implemented infrastructures are briefly reviewed. Next the sets of characteristics identified in the first three sections of chapter 2 are examined in detail. Groups of similar characteristics are identified. The main groups are 'mode of interaction', the 'structure of knowledge', 'presentation', 'governance' and 'technology'. Next the alternatives, which are available for the implementation of each characteristic, are evaluated. The result is a comprehensive matrix of characteristics which can be used for categorizing knowledge infrastructures and which can be consulted for design decisions.

The third chapter addresses a core problem which all scientific knowledge infrastructures face: Finding an adequate structure for representing scientific knowledge within the infrastructure. The problem arises independently of whether a knowledge infrastructure is to be based on paper or information technology. It is complicated by the lack of agreement among scientists as to what constitutes scientific knowledge. Therefore, the purpose of the third chapter is to develop a conceptual model of scientific knowledge which can be used as the basis for the implementation of knowledge infrastructures. First the objectives of developing a structure of scientific knowledge are examined. Then the concept of scientific knowledge is examined from a theoretical and philosophical perspective. This includes an examination of different interpretations of scientific knowledge and of the impact of these conceptualizations on scientific paradigms. Subsequently an object-oriented model of scientific knowledge in the form of a class diagram is developed. Different categories of scientific knowledge are identified by examining the categories which are used by eminent philosophers of science such as Popper, Nagel, Dubin and Bunge and in more practice-oriented text books on research methodologies. The relationship between these categories and their attributes are examined subsequently. This yields a detailed model of scientific knowledge that can be used as the foundation for implementing a knowledge infrastructure.

The last chapter discusses the implications of this research. To show the feasibility of knowledge infrastructures which significantly depart from the traditional model, a system which has been developed by the author and is available on the internet will also be discussed.

2. Characteristics of scientific knowledge infrastructures

Although widespread agreement exists that online knowledge infrastructures will eventually replace the traditional modes of paper-based scientific communication, it is not clear what models will be successful and to what extent online infrastructures will differ from the current system. To answer this question two different approaches can be taken: A deductive approach can be based on a theory of scientific publishing. If the main elements of the current publishing system and their relationships can be identified, then each of the elements and relationships can be analyzed for the impact of information technology. Information technology may lead to the elimination of some existing elements (such as paper documents), to the introduction of new elements and to changes in the relationships and interdependencies (e.g. the sequence of some publishing processes). The theoretical analysis has the advantage that it allows to anticipate developments which currently encounter technological, resource and acceptance problems. Its disadvantage is that it may overlook some impacts and therefore sketch an incorrect vision of future knowledge infrastructures.

The second alternative, therefore, is an empirical analysis of emerging knowledge infrastructures on the internet. In an environment that is not in a steady-state such an analysis has significant limitations. It is unlikely that the current population of knowledge infrastructures is representative of the population of mature scientific knowledge infrastructures. The structure and mix of characteristics may significantly change. However, the set of characteristics which are visible in current knowledge infrastructures may provide an indication of the potential modes in which these infrastructures may emerge. In the following, therefore, first a theoretical analysis of the evolution of knowledge infrastructures will be performed. This will be followed by an empirical analysis. Both theoretical and empirical analysis approaches have limitations. To identify potential additional issues, visions of knowledge infrastructures which have been sketched by renowned scientists will be examined in a third section of this chapter. Finally the potential characteristics of knowledge infrastructures which have been identified will be aggregated into groups. Within each group, the possible variations and issues will then be examined.

2.1. Theoretical analysis

2.1.1. Traditional infrastructures: scientific journals

Paper-based journals are a core element of scientific communication. They emerged in the middle of the 16th century. 'Le Journal des Savans' published in Paris and the 'Philosophical Transactions' published in London are believed to have been the first journals. Today between 80.000 and 100.000 journals are being published worldwide (Tenoir & King 2000). They facilitate the rapid dissemination of new knowledge and create a record of knowledge in a field, a function which they share with scientific books. A scientific (or technical journal) is "a periodical [publication] that an identifiable intellectual community regards as a primary channel for communication of knowledge in its field and as one of the arbitrators for the authenticity or legitimacy of that knowledge." (LaFollette 1992, p 69) Journals and books are not the only means of scientific communication – the importance of direct dialogue among researchers, presentations etc. cannot be underestimated. For the purposes of this analysis, the latter are only relevant insofar as they leave a trace of explicitly, recorded knowledge.

Before a theory of traditional paper-based knowledge can be developed, the desirable characteristics of such a theory need to be identified. Emory (1985, p 30) characterizes a theory as a set of concepts, facts and interrelated statements (primarily axioms and propositions) that are advanced to explain and predict phenomena. Nagel (1961) distinguishes three elements of a theory, an abstract calculus (system of statements), a mechanism for mapping observations to that calculus and an interpretation of the calculus. A more elaborate structure targeted at the social sciences has been developed by Dubin. He distinguishes seven components of a theory (see Table 3 on page 17). A core element of theories are systems of statements or 'models' (Dubin 1978, pp 17-18).

In the natural sciences such systems are usually described in the form of mathematical equations. In the social sciences other approaches including the more ambiguous natural language are frequently utilized. In information systems semi-formal methods have been developed for describing information systems. These approaches lead to models of the complex interplay of people, organizations, processes and information which together constitute an information system. They have proven to be useful for the representation of highly complex information systems - e.g. enterprises supported by enterprise resource planning technology – for which mathematical equations are not suitable. These modeling approaches can be used to formulate scientific theories.

A variety of such modeling methods have been developed in IS. All methods analyze an information system from several different perspectives. Zachmann, for example, distinguished three core perspectives: Information, function and network (Zachmann 1987; Sowa & Zachman 1992). These were modified and extended in subsequent information systems modeling approaches. Scheer (1998b) showed that the network perspective can be treated as special case of an organizational perspective. He also added a process perspective which integrates all other perspectives. An analysis of the publishing system requires at minimum three different perspectives: an organizational perspective which describes the stakeholders and their objectives; an object (or 'information') perspective which describes the primary objects and their relationships and a process perspective which models the sequence of activities (or 'functions') in scientific publishing.

In the following, the traditional publishing system will be analyzed from these three perspectives. The analysis will start with the process perspective because this facilitates the identification of core objects and stakeholders. The object perspective and finally the organization perspective will follow.

2.1.1.1. Process perspective

Although most information systems modeling methodologies include approaches for representing sequences of actions, they adopt a wide variety of different notations. Examples are flow charts, process chain diagrams (eEPC), data flow diagrams and swim lane representations (Scheer 1998b). These notations differ in three main aspects:

1) The range of included concepts. While some methods solely focus on actions, other methods include actors, events, locations, objects etc.
2) Sophistication of conditional branches and looping constructs. While some methods only distinguish sequence and branch, others introduce fine distinctions between conditional branches (e.g. or, xor, and etc.) as well as constructs for loops.
3) Graphical layout: Free-form approaches can be distinguished from semi-structured approaches such as swim-lane representations.

A theory of publishing cannot restrict the range of concepts in a process description to actions. At minimum actors and the objects on which the actions are performed need to be included. Because of the high level of abstraction, detailed semantics are not necessary for sequence and branching operators. It is sufficient to use a generic sequence operator (arrow) which describes that one action is a prerequisite for another as well as the most generic form of a branching operator. Because of the ability to transform a process model from one layout arrangement to another, the graphical layout is not crucial for the theory. In the following the swimlane layout is chosen because of its high clarity.

The traditional publishing process has received much attention in the literature. Menard (1971 p 87), for example, describes it in the context of an analysis of the growth of scientific literature. O'Connor (1986, p 16) embeds the functions of copy-editing in a flowchart of the publishing process. Braman (1995, p 43) identifies information creation, processing, distribution, use and seeking as the main subprocesses of publishing. Spring (1991, pp 9f) distinguishes creation, dissemination, storage and retrieval. Garvey and Griffith (1972) expand the stages of the publishing process by considering indexes and preprints. These descriptions can be used as the basis for the development of a process model (see Figure 2):

The publishing process begins with a researcher or team of researchers who conduct research. At the end of this activity (and sometimes in parallel) a manuscript is prepared which describes the research problem, approach and findings. Upon completion of the manuscript, a paper copy is submitted to a journal. The editor reviews the article and assigns reviewers who evaluate and critique the article. Next, the results of the review process are communicated to the author. If revisions are needed, the author makes changes to the article. This may involve additional research. Once the revision is completed, the revised manuscript is submitted to the editor who forwards it to the reviewers. This cycle continues until the article passes the review, the author gives up, or the article is rejected. After acceptance, the manuscript is ready for copy-editing which is a prerequisite for printing. Once all articles for an issue have been accepted and passed copy-editing, the issue can be printed. Then the issues are distributed to libraries, other subscribers, and in some cases to bookstores, newsstands etc. Finally the articles are available for reading.

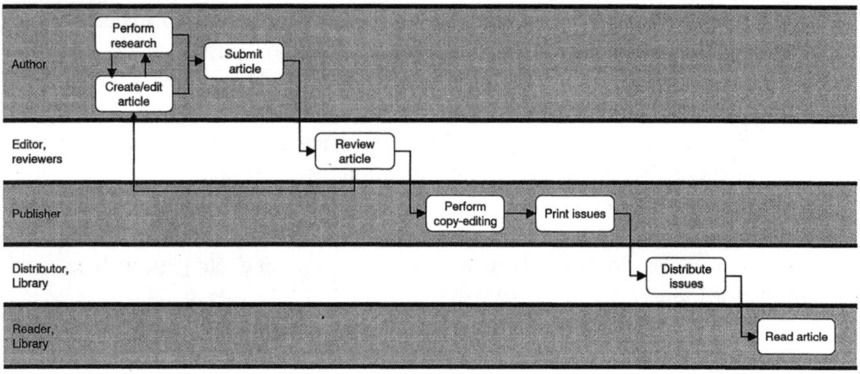

Figure 2. Traditional publishing process

The process is lengthy. It frequently takes more than a year from submission to publishing. This description of the process leaves out some variants and minor issues. For example, editors sometimes solicit articles; authors are typically asked to review the product of copy-editing. In addition it excludes processes which are part of the publishing system but which are less central such as shelving in libraries and subscription handling.

2.1.1.2. Object perspective

This perspective describes a static aspect of the publishing system: the classes of objects which are relevant for it as well as their attributes and relationships. The predominant methods for this perspective are Chen's entity-relationship model (Chen 1976) and its extensions including object diagrams (Booch, Jacobson & Rumbaugh 1998; Fowler & Scott1997). The main elements of such diagrams are shown as rectangles. They represent classes of objects (or: 'entity types', 'categories'). Different types of relationships are shown as lines. An arrow indicates a generalization relationship; a diamond indicates aggregation.

The process model can be examined to identify relevant classes. The publishing process contains two core classes: 'article' and 'issue'. Articles are written by authors, then reviewed, copied and distributed. Every manuscript thus eventually leads to hundreds or thousands of printed copies of the article. This requires the distinction between two different interpretations of 'article': On one hand, an article may be a tangible document which a reader checks out in a library. For clarity, this class will be called a 'printed article'. On the other hand, an article may be interpreted as an intangible entity. This abstract 'article' represents the intellectual content which is the basis for all printed articles. It is the target of citations, comments etc. but it is not equivalent to the final manuscript. The latter is physical and is often changed during copy editing. It looses importance once the article has been printed. In library cataloguing, the terms 'work' and 'book' are frequently used for this distinction (Hildredth 1996, pp 86-87). Both types of articles are linked via 'printed-as' relationships (see Figure 3). Another class is the printed manuscript which forms the basis for an article. Printed articles and manuscripts can be generalized into 'paper documents'. Other paper documents are printed issues and printed volumes. Both categories need to be linked to 'original' issue and volume categories. Neither the issue nor the volume category need to be interpreted as tangible; it is sufficient to characterize them as sets of articles or sets of issues.

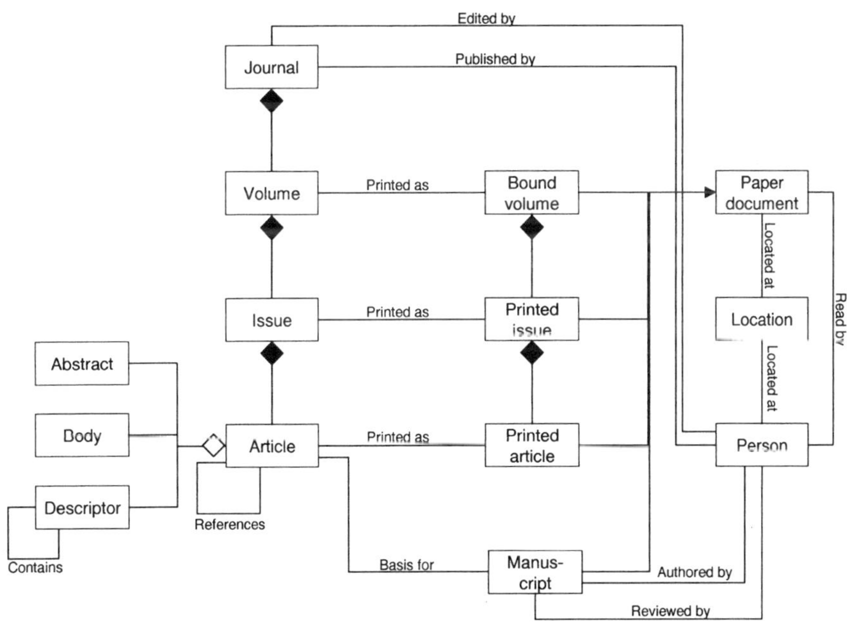

Figure 3. Class diagram for traditional paper-based journals

In the paper environment, reading requires physical proximity between printed document and reader. The reader needs to determine the location of the nearest copy of an article and then has to bridge the distance either by traveling to its location or by having the copy sent to the reader. Thus, location is very important in the traditional environment. In paper-based environments libraries – storage places for print documents – can be very effective in reducing access costs for readers.

Several other classes are relevant for publishing. Articles typically have an abstract and are assigned descriptors or keywords which belongs to some network or hierarchical classification scheme. Articles may reference other articles. References may vary significantly in purpose, ranging from giving credit to related work, criticizing previous work, to providing bibliographic leads (for an overview of different types of references see Case & Higgins 2000). However, these differences are not made explicit in current referencing practice: a single, non-discriminating mechanism is used. Articles appear in a journal. Persons adopt a variety of roles with respect to articles, journals and printed documents. These roles reflect the different categories of stakeholders identified as part of the process model.

2.1.1.3. Organization perspective

2.1.1.3.1. Stakeholders

Seven main categories of stakeholders take part in the publishing process. The first category are authors who perform research, write articles and submit manuscripts. Editors are responsible for a journal; they coordinate the review process. Reviewers ensure the quality of articles. The publisher is responsible for the technical and commercial aspects of publishing. Publishers are usually organizations rather than individuals. Other roles which are associated with the publisher are graphics designers who perform copy-editing, printers etc. These roles will be subsumed under the publisher. Two additional stakeholders are distributors (including the postal service) and libraries which play a central role in facilitating access to scientific knowledge. The final category of stakeholders are readers which are often viewed as the 'consumers' of knowledge. These roles are not exclusive; an individual may simultaneously assume several roles.

2.1.1.3.2. Objectives

In contrast to some other socio-technical systems such as enterprises, no single identifiable entity exists which assumes control over the system's behavior. Therefore it is problematic to ascribe objectives to the overall system. While the publishing system can be investigated as a teleological system and objectives may be used as a metaphor to explain its observable behavior, this behavior ultimately arises from the objectives of the stakeholders and their interactions in combination with constraints imposed by the environment. Therefore it is necessary to examine the objectives of each stakeholder.

Authors may have several objectives. As researchers, they seek to advance knowledge and disseminate their ideas (Lafollette 1992, p 76) to the appropriate audience. From the perspective of dissemination, they view publishing as a giant competition of ideas trying to reach an audience. Journals, hence, are tools for marketing ideas. Another aspect of dissemination is the goal of conserving ideas for posterity. Authors seek recognition and academic rewards in the form of promotion and tenure which are typically tied to publishing (McKnight, Dillon & Shackel 1996, p 352). At present, online journals are perceived as being less respected than print journals. This reduces the incentives for authors to publish in online journals (Butler 1995, pp 175, 168f). Some editors believe that the recognition objective has received too much weight for publishing: "In many ways, publication no longer represents a way of communicating with your scientific peers, but a way to enhance status and accumulate points for promotion and grants." (Hamilton 1990, p 1331)

Editors are often described as the 'gatekeepers of science' (Simon & Fyfe 1994). They are responsible for the scientific aspects of a journal and seek to increase the quality of its articles and the size of its audience. Similarly to authors, they seek recognition from the field and associated academic rewards. Their recognition is strongly dependent on the standing of the journal.

The goals of **reviewers** are more difficult to determine. Reviewers are not compensated for their work and usually remain anonymous. Although they spend a substantial amount of effort for review – typically between 3 and 5 hours for articles in reputed journals (McNutt et al. 1990; Jauch & Wall 1989) – reviewers receive little recognition or academic rewards for their activity. Glogoff suggested that reviewers seek to gain early access to advances in the discipline (Glogoff 1988); Yankauer (1991) emphasized prestige and recognition while other researchers stressed that the primary motivation may be altruistic (Simon, Bakanic & McPhail 1986; Endres & Wearden 1990). Another possible objective would be self interest: Reviewers could increase their own competitive position by raising the barrier on their peers (Gordon 1977; Wilson 1978).

Publishers are interested in the economic success of a journal. Their main goal is to maximize profits. Even non-profit scientific organizations frequently depend on income from journals to cross-subsidize other activities. Additional objectives – if existent – are similar to the objectives of journal editors: Increasing the journal's quality and the size of its audience. Scientific associations often aim to document scientific knowledge for the future (LaFollette 1992, p 75).

Similarly to publishers, **distributors** are primarily interested in the economic aspects of their activities. Besides providing access to knowledge for their constituents, **libraries** strive to preserve and to organize knowledge (Okerson 1996 p 197).

Finally the **reader's** objectives must be considered. Readers seek to stay abreast of new developments in their fields (McKnight, Dillon, & Shackel 1996, p 352). They want to apply the knowledge. Many readers turn to journals to study and assimilate new knowledge. Table 2 shows the objectives of all stakeholders.

These goals are often in conflict. The priorities assigned to each of the goals may differ from publishing system to publishing system. In the current system, the high costs incurred in printing and distribution are a major constraint. This reduces the ability of the individual researcher to preserve his thoughts for posterity or to disseminate his ideas by himself. It increases the weight of the editors quest for quality but may have the negative effect of limiting the generation of new ideas – which is a goal of the authors. Some researchers have argued that the current journal publishing system is not very reader-friendly. Assimilating knowledge about a topic from journals requires substantial effort. In addition, the current priorities may be favorable to the goals of the publishers who have raised revenues considerably in recent years (Cummings et al. 1992). Of course, no single instance exists which assigns these priorities; they become evident during the operation of the system and may change at any time.

Table 2. Objectives of stakeholders in the publishing system

Authors	Reviewers	Editors
Disseminate their ideas Conserve their ideas Create new knowledge Obtain recognition Receive academic rewards	Keep abreast of novel developments Receive academic rewards Enhance their competitive position as authors Minimize effort	Maximize quality of contributions Maximize readership Receive academic rewards
Publishers, Distributors	**Readers**	**Libraries**
Maximize profit	Preserve knowledge Organize knowledge Maximize access to knowledge	Keep abreast of novel developments Assimilate useful knowledge Minimize cost of accessing relevant knowledge

The move towards electronic media may shift the power balance between the different stakeholders and therefore change the priorities of the overall system. The degree to which the shift will occur does not only depend on changed constraints; it strongly depends upon the way that the collective bargaining process will play out in the coming years.

Among these goals, those which are directly related to knowledge stand out. While it is clear that individuals require incentives and rewards, in an ideal publishing system, these mechanisms would have evolved to ensure pursuance of knowledge-related goals. The scientific culture downplays the importance of rewards and recognition. The goal of 'receive academic rewards', for example, is largely an externally imposed goal: academic institutions have chosen to raise the incentives for performing these roles by making these roles a criteria in their evaluation of a researcher's performance. Efficiency issues are relevant for every human activity. The incentive goal, however, should be subsidiary to the overall knowledge-related goals.

2.1.1.4. Validation

The theory presented above and shown in summary form in Table 3 using Dubin's framework is not very controversial. Most of its elements can be mapped directly to observable phenomena. Reprints, manuscripts and journal issues can readily be identified. Journals can be identified via ISSN or by searching any library catalog for journals. The distinction between scientific and non-scientific journals is not as clear-cut (Lafollette 1992, p 69). A journal's mission statement can be used as well as articles which evaluate and rank-order scientific journals.

Table 3. Theory of traditional scientific journals following Dubin's framework

#	Component	Description	Traditional journal
1	Unit	Concepts (real and nominal) which are at the core of the theory	Components of object model (10), process model (8), stakeholders (5)
2	Law of interaction	Statements which formally describe the interaction	Relationships between entities, process flow
3	Boundaries	Describes the (external) context in which the laws of interaction hold	Scientific paper-based journals
4	System state	States within which the units interact differently	Theory assumes the steady-state.
5	Proposition	A truth statement (usually a subset or consequence of the laws of interaction)	Journals are organized by issue Review always precedes printing Distribution precedes reading
6	Empirical indicators	Empirical indicators related to terms in the propositions	Average number of articles per issue of a journal Number of scientific journals which are organized by issue – not by article or volume or by another mode Percentage of scientific journals which publish articles without any form of review
7	Testable hypothesis	Translation of the proposition into testable hypothesis	No scientific journal exists which publishes articles without reviewing them at all The number of articles per issue is always larger than one

An approach for validating the object model would be to compare it to an electronic library system. This would confirm the distinction between physical documents and their abstract counterparts which are often called 'titles'. Library systems often do not contain a 'location' entity; to find a document, the reader needs to consult a floor plan which maps document identifiers to locations. This confirms that location is relevant, even if it is not directly shown on a library card or in an electronic library record. The location of a person is not recorded in library systems; however, persons always have a specific location. In a paper environment, costs for reading are determined, in part by the distance between the next copy of the article and the location of the person.

The process model is highly abstract and is confirmed by accounts in various research articles (Menard 1971, p 87; LaFollette 1992, p 75; O'Connor 1996, p 16). However, the model does not cover all processes which are relevant for publishing, e.g. reviewer selection, subscription and marketing processes, etc.

Therefore the model is only a partial model. The set of stakeholders can be cross-validated by the publishing process. Some important stakeholders may be missing, however. Taken to the extremes, one could argue that everybody, in particular government, research agencies, teachers, entrepreneurs etc. are affected by the publication system. Their elimination from the system reflects their limited direct involvement in the publishing process. Overall, therefore, this model rests on solid ground.

The theory can be used to explain some characteristics of the current publishing system. The fact that articles are bundled into issues, for example, is a consequence of the economics of publishing. High fixed costs in the printing and distribution phases make bundling more cost-effective than individual printing and distribution of each article. The theory can also be used to derive propositions and hypotheses which can be supported empirically. It follows from the theory, for example, that journals do not publish articles without any form of review. Another consequence of the theory is that articles need to be distributed before they can be read. Of course, few researchers would doubt these hypotheses. The risk of falsification and therefore the theory's value seem to be low. The picture changes, however, once a shift of media is considered. The hypotheses become much more valuable, if they are not universal to all publishing systems but only apply in the traditional environment. This will be shown in the next section.

2.1.2. Impact of information technology

Information technology leads to a variety of potential changes in scholarly communication. The theory of traditional journals can be used as the foundation for examining these changes. First, each component of the traditional theory needs to be analyzed for potential impacts. Subsequently, the interdependencies between the changes need to be considered. Which of the potential changes will actually be implemented depends on many soft factors and therefore is difficult to anticipate. As a consequence, the following simplifying assumptions need to be made about the mechanisms which determine the evolution of online journals:

Assumption #1 (Utility): The evolution of online journals will increase the utility of journals for their stakeholders. Thus, when the overall benefits of a feature to the stakeholders are larger than their overall costs, then this feature will be adopted.

Assumption #2 (Selective activation): This assumption is a consequence of the utility assumption. Information technology enables features which may be selectively turned on or off by each user. If the evolution of online journals increases the overall utility of their stakeholders, then features which can be selectively activated will be adopted provided that there are some users who benefit from this feature and that the overall costs are lower than the benefits to this group of users.

Assumption #3 (Legacy): Some features of traditional journals are known to be a consequence of limitations of the paper medium. The evolution of online journals will not replicate these features in mature online journals if the limitations no longer persist.

These assumptions disregard some of the issues on which the evolution of online journals may depend. For example, they do not consider power differentials and governance structures which are difficult to evaluate systematically (Hurd 2000). Nevertheless, the theory is valuable to determine possible futures for scientific knowledge infrastructures and to better understand their observed evolution.

In the following, each component of the traditional publishing system will be examined. The most obvious change occurs in the object perspective where electronic documents appear and where print documents might be eliminated as a consequence. Therefore the analysis will begin with this perspective. The process and organization perspectives will follow. Finally, the cumulative effect of the individual changes in all perspectives will be evaluated. A summary of the elements of the theory of online journals to be developed subsequently is shown in Table 4. Table 5 on page 20 lists key propositions about the differences between online and traditional journals.

Table 4. Theory of online journals

#	Component	Description	Online journal
1	Unit	Concepts (real and nominal) which are at the core of the theory	Components of object model, process model, stakeholders
2	Law of interaction	Statements which formally describe the interaction	Relationships between entities, process flow
3	Boundaries	Describes the (external) context in which the laws of interaction hold	Academic online journals Assumptions 1-4
4	System state	States within which the units interact differently	States other than steady state
5	Proposition	A truth statement (usually a subset or consequence of the laws of interaction)	Propositions 1-11
6	Empirical indicators	Empirical indicators related to terms in the propositions	e.g. prop#2: ratio of online journals which allow authors to change their article after submission. prop#6: ratio of online journals which distinguish issues in their table of contents.
7	Testable hypothesis	Translation of the proposition into testable hypothesis	Propositions 1-11

2.1.2.1. Object perspective

The main change which occurs in this perspective is the introduction of an 'electronic document' which can support multiple media in an integrated fashion. "For the first time in the history we are faced with a unitary form of record that is suitable to all forms of human expression" (Bizjak 2000). The electronic document exhibits novel characteristics such as the ability to change its contents, to alter its presentation etc. In addition, the relationships between an electronic document and other elements of the object perspective such as articles, printed documents etc. need to be analyzed. In the following major differences between online and traditional journals will be identified as numbered propositions (see Table 5).

Table 5. Impacts of information technology on online journals

#	Proposition	Argument	Theory element
1	Electronic documents can be changed after publication.	Technical	Object model: property of electronic document
1a	In an online journal, publishing is reversible.	Derived	Proposition 1
2	Mature online journals will allow authors to update articles	Utility (Ass#1)	Proposition 1
3	An electronic document can be presented in different ways to different readers	Utility (Ass#1)	Object model: property of electronic document;
4	On the WWW the physical location of article and reader are irrelevant.	Technical	Object model: location. Process model: distribution
4a	If a document consists of several parts, it is irrelevant where each part is located.	Derived	Proposition 4
5	In an online journal, all readers share the same document – no more copies	Technical	Process model: printing & distribution; object model: printed-as relationship
5a	On the WWW, there is no need to keep copies of a publicly accessible document. Links to the original are preferable.	Derived	Proposition 5
6	Journals will cease to be organized by issue	Legacy (Ass#3)	Process model: printing & distribution; object model: Aggregation relationship article→issue
7	On the WWW, scientists can self-publish any article	Factual	Process model: printing & distribution; impact 3
7a	Publishing will cease to be an important function of e-journals	Derived	Proposition 7
8	An online journal may accept submissions from persons who are not the author of the article	Technical	Proposition 7, proposition 5a,
9	Reviews will be multidimensional, not binary	Selective activation (Ass#3)	Process model: review, reading; object model: descriptor, preferences.
10	Articles will be published first and reviewed thereafter	Selective activation (Ass#3)	Process model: review
11	An article may appear simultaneously in several journals	Utility (Ass#1)	Props. 4 and 8; multiplicity of aggregated relationship article→journal

An important characteristic which sets electronic documents apart from paper documents is the ability to update their contents. This ability is a consequence of hard drive technology which allows to erase and rewrite information repeatedly. Thus it is technically feasible to update online publications as new findings emerge, as errors are uncovered or as new relationships are identified. This leads to **proposition #1**: *Electronic documents can be changed after publication.*

Of course, technical feasibility does not imply actual adoption by online journals. Most current online journals, for example, do not allow authors to change their articles. Nevertheless, exceptions exist: MISQ Discovery explicitly distinguishes between 'archival' and 'live' articles. The latter are kept up-to date by the authors (www.misq.org/discovery/about.html, 9/17/2002). Similarly, Communications of the AIS (cais.aisnet.org), a new online journal, allows limited updates by the authors.

The ability to change articles in online journals is a significant departure from current practice. A cost-benefits analysis can yield some indication of whether it will be adopted. From the publisher's perspective, the marginal costs of updating an article are almost zero. Once an access management system (and potentially a version management system) are in place which require a limited one-time investment, authors can directly access and modify their articles with negligible additional cost for the publisher. Thus even small increases in the value of the knowledge reflected in an article would ensure that marginal benefits exceed marginal costs. If the author is included in the analysis it is sufficient to assume that the author would make a correction only if the value of the correction is higher than his cost of making the change.

One might argue that additional costs may be incurred if the quality of articles suffers because authors use this mechanism to circumvent review. This argument has several flaws. First, the ability to update does not imply that updates remain unreviewed. In the case of the Stanford Encyclopedia of Philosophy (plato.stanford.edu), for example, reviewers are automatically notified whenever an author updates an article. Second, even if changes were not reviewed this does not necessarily imply a reduction in quality. When authors are rewarded by number of publications they have an incentive to publish new findings in new (reviewed) articles rather than to significantly extend existing articles and circumvent review. They are more likely to add clarifications and to correct grave errors discovered after publication. In balance, the ability to update will increase rather than decrease quality. Moreover, if a version management system is in place, readers can be given the option of selecting the version – the reviewed original or one of its un-reviewed successors – which they want to access. Thus each reader can match the risk of lower quality content with her needs of accessing current knowledge. The benefits of incorporating error corrections, clarifications and results of suggestions from readers into an article after publishing will be higher than the costs. As a consequence, online journals which seek to maximize the value to their readership, will allow articles to be changed: *Mature online journals enable authors to update their articles* (**proposition #2**).

Proposition #1 has another immediate consequence. It is possible to withdraw a published document from the scientific body of knowledge. In contrast to the print medium, therefore, *in an online journal, publishing is reversible* (**Corollary #1a**).

In contrast to paper, electronic documents can present their content in different ways. They are not limited to a single, predefined form of presentation. For example, data can be shown as diagrams or tables. Detail can be hidden from the casual reader but shown to the expert. Formulas may become active, allowing the reader to enter her own data and to perform instant calculations. In an electronic environment content and presentation need to be clearly distinguished from one another. This leads to **proposition #3**: *An electronic document can be presented in different ways to different readers.*

This has significant implications for the design of electronic journals: When a reader opens a paper journal he is presented with the same table of contents and the same articles as every other reader. In an electronic environment, the presentation of a journal or article may be dynamically adapted to the reader's needs. The personalization of journals also has implications for the selection of articles; they no longer need to appeal to the 'general audience' to be acceptable but can be targeted at specific sub-communities. This capability can be reflected in the object perspective by extending the 'descriptor' category. A relationship 'preference' can be added between 'person' and 'descriptor'. This enables an online journal to serve only those articles to a reader in which he is interested. The 'preference' relationship is not the only way in which preferences can be recorded. Preferences can also be recorded directly as attributes of the 'person' class.

The internet revolutionizes access to information. Information is available instantly almost anywhere in the world at costs which do not depend on the distance between article and reader. While physical location matters on the network level, it becomes transparent to the reader. In addition, the mechanism of uniform resource locators ensures that local and external files can be treated in the same way. Therefore, publisher and reader do not need to handle local documents differently from documents physically stored on another continent. The consequence for the object model is the elimination of the location entity. While location remains relevant on the physical level, the stakeholders of the publishing system no longer need to be concerned about it. This leads to **proposition #4**: *On the WWW the physical location of article and reader are irrelevant.*

In practice, the ideal of complete location independence has not yet been achieved: Response times still depend somewhat on the distance between reader and article. The practice of establishing mirror sites on different continents shows that some location dependence remains. Further evolution of the internet, however, will gradually reduce these problems.

Proposition 4 immediately leads to the following corollary:

Corollary #4a: *If a document consists of several parts, it is irrelevant where each part is located.*

This corollary is a direct consequence of the location-independence of the internet. If a document consists of several sections, it is irrelevant for the reader whether some of the sections are retrieved locally or from a distant location.

In the traditional model, a distinction is made between an 'article' and the copy of the article which is available to the reader. The distinction is important as there are many activities which specifically relate to the copies. They are printed, distributed, catalogued in libraries, shelved, checked-out and checked back in, receive annotations etc. Thus each copy assumes an independent existence.

It is questionable whether the same phenomenon applies to electronic documents. Because of proposition #4 the location of an article is irrelevant on the internet. Thus there is no need to store copies of an article locally or in libraries to simplify access. Information technology allows quasi-parallel access to the same physical file – the 'original' – by several readers. The mechanism for retrieving an article is transparent to the user. But one might argue that the retrieval of a file by an internet browser is a copying process: a duplicate of the file is created and usually held in the cache of the browser. However, the duplicate is transient; it is only stored for a short time. The next time that the reader opens the same article, the browser will first access the original to determine whether the file available in its cache is still current. Only if this is the case, will the cached document be used. To the reader, the process is transparent; he is assured of always retrieving the latest version of the original. Thus online retrieval of a document is not similar to a printing or copying process The analogy with print copies faces a second problem: When a copy of an article is printed, the copy always contains the full article. In contrast, electronic documents may be retrieved in parts. If a document consists of several components, only those sections which are currently being read will be retrieved. As a consequence, the distinction between original and copy looses its relevance in the electronic environment. Thus, *in an online journal, all readers share the original.* (**Proposition #5**).

The difference is especially important when articles change and for capturing annotations. In the print environment annotations can only be made locally to the copy of an article. The internet provides the potential of centrally capturing all annotations to a document. This does not require that each annotation has to be public. Therefore no separate 'electronic document' class is needed in the class diagram. Without version management, the file corresponding to an article can be treated as an attribute of 'article'. With version management, a 'version' relationship needs to be added between articles.

A consequence of propositions 4 and 5 is the following corollary:

Corollary #5a: *On the WWW, there is no need to keep copies of a publicly accessible document. Links to the original are preferable.*

Having concluded that it is not necessary to add a new 'electronic document' category to the class diagram, the question remains what will happen with the 'printed document' category. The emergence of electronic documents does not automatically imply that printed documents will become obsolete. Many online journals are hybrid, publishing both a printed and an online version. Furthermore, the current display technology is widely considered inadequate for online-only viewing:

> "The accumulated evidence suggests that scholars … much prefer to read printed materials than screens, and much prefer the portability, browsability, and other familiar characteristics of a book or journal to the electronic formats now

> available." (Riecken 1989; see also Kling & Lamb 1996, p 17; McKnight, Dillon & Shackle 1996, p 354)

If readers and reviewers prefer printouts to reading from a display, then this may decentralize printing but it does not eliminate paper documents.

The problem of how to represent printed documents in the class diagram may be treated in one of two ways: one may argue that paper documents will persist in the online environment and therefore have to be reflected in the theory. Alternatively, one may argue that it is just a question of time until display technology will became mature. The first approach has the advantage of reflecting the current situation and the disadvantage of leading to a hybrid paper-IT theory. One might expect that centralized printing and distribution may at times be more economical than decentralized printing. In addition, the degree of user preference for paper vs. electronic displays would become a key factor of the system. Unfortunately, this factor is difficult to predict and likely to change with advances in technology.

The second approach has the advantage that a conceptually clear analysis can be presented which is not limited by problems of transition from one technology to another. The drawback that the system does not reflect current technology is offset by the advantage that the designers of publishing systems can build their systems to take into account structures towards which online journals may converge in the future. Moreover, display technology is maturing. While CRT monitors have significant disadvantages in size, weight and energy consumption, - and therefore it is no surprise that the decades-old visions of the paperless office have not yet become reality – more promising alternatives are emerging. Prices of flat panel monitors are falling significantly and may soon reach a point which allows widespread adoption. Recent breakthroughs by IBM and XEROX have lead to flat panels with a resolution that is similar to paper. In addition, electronic books based on flat panel technology are now being marketed which meet size and weight constraints of conventional books. Furthermore electronic books have significant advantages over paper documents: They can store large collections of documents, they allow to organize and annotate files systematically and they can provide added features such as search and analysis. Flat-panel technology, therefore has the potential of replacing paper although it still has the disadvantages of increased weight, power, cost, etc. Another display technology which is currently being developed at XEROX Parc, MIT and their spin-off companies promises to overcome these limitations: Both institutions are developing a high tech alternative to traditional ink by using small magnetically charged color particles which are black on one side and white on the other. An electric field allows to reorient the micro-encapsulated particles and thus 'print' an image on paper. This technology promises very low power consumption. Flexible displays with a resolution of 200 DPI which are only slightly thicker than paper have already been demonstrated (Jacobson et al. 1998; Daviss 1999; Mann 2001). Therefore it is not unrealistic to assume that at some point in the future, paper will cease to provide a significant advantage over other display technologies. Therefore, further analysis of online journals will be based on the following assumption:

Assumption 4: *Display technology will be sufficiently mature to replace paper.*

As a consequence, the 'printed document' entity and its sub-entities can be removed from the object model. Among the remaining entities, issue and volume will be reviewed below in the context of the printing, distribution and copyediting processes. The journal entity persists: access rights, review procedures and physical storage are usually handled on this level. Small changes are required for the author, review and read relationships which were moved from printed documents to the article entity. The resulting class diagram is shown in Figure 4.

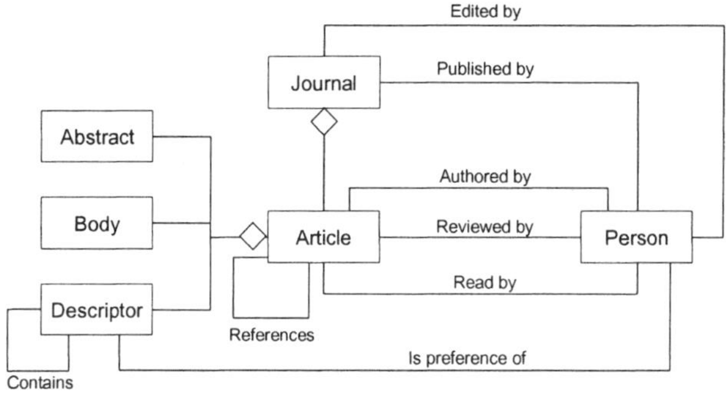

Figure 4. Reengineered object model

2.1.2.2. Process perspective

The elimination of printed documents and article copies has significant implications for the publishing process. The printing and distribution processes will be most directly affected and therefore will be examined first. Then the remaining phases of the process and its sequential structure will be reviewed.

The elimination of paper copies significantly changes the cost structure of publishing. As all readers share the original document (proposition #5), online journals do not need an equivalent to printing. The distribution process also changes. In online journals, articles are distributed on demand and the distribution mechanism is embedded in the fabric of the internet: It provides access to the document whenever a reader requests an article. In contrast to the paper environment, where distribution was a prerequisite for reading, distribution now becomes an aspect of reading. It is carried out in parallel with reading. This is particularly evident when documents consist of multiple parts. Only the first part will be downloaded when reading begins. As the reader reaches a new section, the corresponding file is retrieved via the internet. Thus electronic documents will often only be distributed partially.

The elimination of the printing and the merging of distribution and reading have the potential of significantly shortening the publishing process. These are not the only implications, however. The traditional process contains structural breaks within the copy editing task as well as between distribution and reading. The object on which copy editing is performed, is an article or an issue. Its predecessor (reviewing) focuses on articles. The object which is printed and distributed, however, is not the article but an issue - a collection of articles. The last task – reading – again focuses on the article. These structural breaks are not an accident. They reflect large fixed cost blocks in distribution and printing. The mailing costs for a single article, for example, do not differ greatly from the mailing costs for an issue. Economic necessity, therefore, has led to the bundling of articles into issues which then is reflected in the process and object models.

Information technology changes the cost structure. Printing costs can be eliminated (Guedon 1995, p 69). On the internet, bandwidth – the capacity to transmit certain amounts of information per time interval – is a main cost block. The actual transmission typically does not lead to additional costs although internet access providers might adopt variable pricing schemes in the future. As a consequence the individual distribution of articles is more cost-effective than the distribution of whole issues (McKnight, Dillon & Shackle 1996, p 357). The transmission of issues has significant disadvantages in the case of variable pricing; even in the case of zero transmission costs the transmission of issues would unnecessarily increase bandwidth. Therefore, in the absence of the economic need, journals will abandon the organization by issue (Cummings 1992, p 131). This leads to **proposition #6**: *Journals will cease to be organized by issue.*

This trend is already visible in several current electronic journals which have begun a practice called 'pre-publishing' where an article is made available before the issue has been officially released. In addition, some electronic journals such as the journals of the Association of Information Systems and the Journal of Artificial Intelligence Research have eliminated the concept of issues. A similar argument holds for volumes. Volumes are needed to reduce costs of search and handling in a library. Retrieving an article from a shelf which consists of individual issues is more difficult than retrieving the same article from a shelf which contains bound volumes of the same journal. The resulting reduction of the object model is shown in Figure 4. Even after the elimination of both constructs, information about volumes can be derived from the publishing date of an article. Sometimes issues are organized around a topic; this is still possible in the revised structure shown in Figure 4 via a mapping to descriptors (e.g. keywords and subject areas). In a paper environment, information about volume and issue are needed to physically access an article. Thus the time of publishing serves as an indicator of the article's location. On the world wide web, however, the location of an article ceases to be of importance (proposition #4). Therefore volume and issue information loose much of their importance. Certainly, every article will have an attribute denoting the publishing date; however, access to the article may be provided via different mechanisms, e.g. through search engines, via subject area references or through other classification schemes. Each article can be published when ready. The binding of articles into issues is natural to paper; in an electronic environment, it has no special meaning.

In the paper environment, copy-editing is a prerequisite for printing. It prepares a manuscript for typesetting and may also improve language, comprehensibility, consistency and presentation of the work (O'Connor 1986). In the electronic environment the author already provides the manuscript in a format which is suitable for distribution or which can be converted automatically. Therefore, direct publishing without copy-editing is technically possible. Researchers disagree about the necessity of the aesthetic part of copy-editing. While O'Connor (1986, p 6) argues that "it is copy-editors who defend readers from the worst that authors can do," others argue that they have little impact on an article's content. With the increase of scientific publications which are submitted in camera-ready format, there can be little doubt that it is feasible to operate an online journal without copy-editing. Therefore copy-editing ceases to be a necessary activity.

The cost structure of online publishing dramatically decreases the entry barriers for publishers. All that is needed is an internet connection and space on a web server. As the large number of electronic journals shows, new journals can be founded rapidly and without significant risks. Moreover, the vast majority of scientists have personal internet accounts which enable them to self-publish their articles (Armstrong 1997, p 77; similarly: Logoze & Fielding 1998). This is in strong contrast to the paper environment where researchers were dependent upon publishers for the distribution of their articles. For scientists, publishing space on the internet is not a scarce resource: *On the internet, scientists are able to self-publish any article.* (**Proposition #7**)

This implies **Corollary #7a:** Publishing will cease to be an important function of electronic journals.

Scientists may still prefer journals to market their ideas – although they may find that other mechanisms including search engines may be attractive alternatives. But the value-added which journals provide is not publishing. If publishing space is not scarce, then an electronic journal cannot add value to an author by providing access to his article. As the physical location of an article no longer matters, journals are better envisioned as entities which publish links to self-published originals. As long as the access to scientific knowledge is free, there is no difference whether a journal article is physically stored in some web server that belongs to the journal publisher or in any other place on the web such as the author's homepage.

In the case of writing/editing articles, the switch towards information technology has already occurred in practice. Most researchers now write their articles on a word processor and store them in electronic form. A larger change occurs in the submission phase. Traditionally manuscripts are submitted on paper. In an electronic environment it is possible to send the electronic version directly to an editor. In addition, this electronic version can adopt a format which is ready for publishing. The next generation of word processors will further change the picture of online submission. Upcoming word processors, for example, will enable an author to directly edit his document in HTML format on the internet or an intranet. As the location of an article is irrelevant, a manuscript may never need to change its physical location during the whole publishing process. Once the manuscript is completed the author may set a flag on his server to change the publishing status and to allow external access. He can then inform the editor of a journal about its location. Once the article is accepted, the journal only needs to publish the link to make the article available to its readership. Submission thus can be reduced to an email which informs the editor about availability of a new article on the internet and its location.

This variation in the process introduces another significant change: if the author has already self-published his article on the internet (proposition #7), then the article is already available on the internet. It can be incorporated into any existing journal through a link (proposition #5a) without explicit consent by the author. This changes the set of persons who are able to 'submit' an article to a journal. In the traditional environment, the only person who can submit an article is its author. In the online environment, once an article has been made publicly available somewhere on the internet, any person can submit this article for 'inclusion' into a journal. Technically this only requires a link to an already published article and therefore does not constitute copyright infringement. Instead of waiting for a submission, journals may search the web for excellent articles which fit the editorial mission. This leads to **proposition #8**: *An online journal may accept submissions from persons who are not the author of the article.*

The next aspect of the traditional process which needs to be analyzed is review. Even in the traditional environment, the review process has been the subject of heated discussion. There is widespread consensus that mechanisms are needed which ensure the quality of scientific knowledge. Scientists cultivate a skeptic position towards knowledge and strive to accept only what can be clearly and rationally demonstrated. This position has been institutionalized through the process of peer review where manuscripts are reviewed before publication for their quality. The review process screens papers and ideally ensures that only articles of high quality – which are worthy of the attention of the research community – reach publication. Many different forms of peer review exist. Most – but not all – are blind: the reviewer is not informed about the author. Ziman (1968, p 11) wrote that "that the referee is the linchpin about which the whole business of science is pivoted." Reviewers are often seen as the "gatekeepers of science." (Beyer 1978)

On the other hand, the current review process has been widely criticized. In a detailed review of the literature on peer review, Campanario (1998a, 1998b) identifies the following criticisms: Several studies have examined the reliability of review and discovered that reviewers frequently disagree in their observations and conclusions. This is aggravated by a lack of standards for review (Sternberg 1985). A second problem is the accuracy of review. Several researchers have found that reviewers often fail to detect key mistakes in a paper (Murray 1988; Campanario 1998a, p 197). A third criticism of peer review is that the system may be biased towards reporting only positive results. While some authors regard this as an advantage, other researchers may study the same issue and come to the same conclusion – thus unnecessarily duplicating efforts. The peer review system has also been criticized to be biased against replication. The editorial policies of many journals require novel and original content. Several studies have shown that replication studies are less regarded in the social sciences than in the natural sciences (Bornstein 1990; Mahoney 1985). Furthermore, blind review has been criticized as ineffective in many cases: reviewers with good knowledge of the topic are often capable of identifying the author even in blind review. Campanario (1998b) identified additional problems such as reviewer negligence, favoritism and the existence of 'invisible colleges'.

Several studies have empirically examined problems of peer review. At the end of the Seventies, Peters & Ceci (1982) tested the peer review practices of psychological journals. They resubmitted 12 articles which had recently been published to the same journal. The delay was between 18 and 34 months. Before submission, the names and affiliations of the authors which originally held significant prestige were replaced with fictitious names and affiliations that indicated little research prestige. Of the nine articles reviewed, eight were rejected and only one was accepted. While Peters & Ceci regard this as evidence of reviewer bias, other researchers have pointed out that a random review process with high rejection rates would also explain their observations (Beyer 1982).

More importantly, reviewers are often criticized for rejecting brilliant work. Campanario (1998a) reports several studies and cases in which articles that later were highly cited had been initially rejected. Similarly Horrobin argued that the function of the peer review is "to identify and reject poor work, to improve and accept good work and to let the best through unimpeded." (Horrobin 1982, p 33) While the system is successful with the first two functions, peer review fails with respect to the last. He argues that it is a "myth" "that all scientists are peers – that is people who are roughly equal in ability." (Horrobin 1982), p 34) In addition, reviewers typically do not recognize exceptional ideas and reject them. Ellison (2000) has developed a dynamic equilibrium model of the review process. It shows that review norms evolve to focus on craftsmanship and polish rather than importance and originality. Such criticisms have culminated in harsh conclusions such as:

> "The *concept* of peer review is faulty at its core" (Horrobin 1982, p 33)
>
> "Peer review favors unadventurous nibbling at the margin of truth rather than quantum leaps." (Lock 1985)

In the light of such criticism the transition towards online knowledge infrastructures presents an opportunity for significant improvements to the traditional review process. Ginsparg (1996) has argued that "the current implementation of peer review, as employed by paper journals, needs to be entirely rethought in view of new possibilities afforded by electronic publication and dissemination." Varian (1997) has identified several of these potential changes:

> "When publication was a costly activity, it was appropriate to evaluate papers prior to publication. Ideally only interesting and correct work manuscripts would undergo the expensive transformation of publication. Furthermore publication is a binary signal: either a manuscript is published or not. Electronic publication is cheap. Essentially everything should be published, in the sense of being made available for download. The filtering process will take place ex post, so as to help users determine which articles are worth downloading and reading. As indicated above, the existing peer review system could simply be translated to this new medium. But the electronic media offer possibilities not easily accomplished in print media. Other models of filtering may be more effective and efficient."

One of the key issues he raises is the binary nature of the traditional review process. Review requires a detailed evaluation of the article, however, its main outcome is the acceptance or rejection decision. The binary focus of review is a consequence of the need to assign scarce publishing space. Although more detailed information about the evaluation might be useful to readers, space limitations prevent it from being published together with an article.

This situation differs substantially in an online environment. As all articles can be published (proposition #7), the review process does not need to resolve conflicts about access to publishing space. Furthermore, publishing space is not scarce and review information can be attached to an article. This information can be published in unstructured form by attaching the reviewers' comments. Moreover, the information may be aggregated into ratings along various quality dimensions. For example, each unit of knowledge may have at minimum two quality attributes which describe their degree of rigor and relevance. "The binary division between published and unpublished data can, in principle, be replaced with a spectrum." (Hilgartner 1995, p 247) Articles can easily be ranked according to different evaluation criteria, e.g. rigor, relevance to specific target audiences etc.

A multi-dimensional rating mechanism has several advantages. It shifts the reviewers' task from making an acceptance or rejection decision to characterizing the advantages and problems of an article. It can be left to each reader how to aggregate the criteria into a personal acceptance/rejection function. A practitioner, for example, might favor articles which are high in relevance while a Ph.D. student might prefer rigor. Technically, this can be achieved by adding filtering functionality to a knowledge infrastructures. Such functionality allows readers to define filters based on a combination of quality criteria determined in the review process, subject area descriptors, read or citation counts etc.

These filtering mechanisms then enable readers to selectively access those papers they are interested in. This may be the ten papers in IS which have received the highest overall ratings. It can also be all papers in a specific subject area which exceed certain thresholds in rigor and relevance. Many other combinations are possible. The separation of evaluation and publishing decision – which is economically feasible on the internet – has the advantage that review can provide more specific information about a contribution than can be conveyed through an acceptance/rejection decision. It allows to make additional distinctions among high quality papers. Furthermore, it allows accepting articles even if they are of interest only to a small subset of the audience. This approach is preferable over the current practice of significantly raising the bar for such articles (Whetten 1989, p 495). While it will depend on many factors how quickly electronic journals will adopt more refined review and associated filtering schemes, there can be little doubt that on the internet binary review decisions are sub-optimal (Ginsparg 1996). This leads to **proposition #9:** *Reviews will be multidimensional, not binary.*

After reviewing the individual phases of the publishing process, the impact on the overall structure needs to be examined. In the traditional environment, the sequence is obvious: review must occur before printing which occurs before distribution. This, in turn, is a prerequisite for reading. In an online environment, the printing function is eliminated and the distribution function becomes an aspect of reading. The process might be reduced to the sequence *review* → *read*. In a paper environment, there are no alternatives to this sequence. Because publishing space is scarce, a review mechanism is needed to fairly allocates publishing space (LaFollette 1992, p 120) and to let those articles with the highest value pass. In the online environment publishing space is not scarce (proposition #7) and articles can be changed after publication (proposition #2). It is technically possible to completely remove an article from a journal after it has been published (proposition #1a). As a consequence, it is possible to publish an article first and review it thereafter. While this is a significant departure from accepted practice, instant publishing has significant advantages: it reduces cycle time significantly by parallelizing previously sequential processes. The counter-argument that this might lead to a flood of low-quality articles incorrectly assumes that the reader is not able to distinguish between unreviewed and accepted articles. In addition, in an electronic environment, filters may be applied which can selectively eliminate articles. Filtering technology can be used to provide every article – including unreviewed articles – to a researcher in his area of expertise while eliminating unreviewed articles when the same researcher browses other areas.

As a consequence, what is obvious in the traditional environment (that an article needs to be reviewed before being printed and distributed) is much less obvious in an electronic environment. Current practice shows that some knowledge infrastructures have adopted instant publishing: Ginsparg's physics preprint archive (xxx.lanl.gov) is one of the most successful academic sites on the internet. It provides access to preprints – articles which have been submitted to a physics journal. The success of the physics preprint archive has led to the development of several other preprint archives in other areas of science. This leads to **Proposition #10**: *Articles will be published first and reviewed thereafter.*

Note that this conclusion does not depend on the assumption that internet publishing is free, or that quality is not important; it is merely based on the assumption that a) the publishing decision is reversible and b) that a reader can distinguish whether an article has been reviewed or not.

Because of proposition #4, the storage location of an article is not relevant on the internet. Because of proposition #8 journals will no longer provide publishing as their main service. If a journal incorporates articles which are not stored under its direct physical control, then one article may be published in several journals at the same time. While this would have been economically inefficient in the traditional environment, it is efficient in an online environment because no duplication occurs; journals can share an article by pointing to the same URL.

It may become inefficient, however, if several journals review the same article. Therefore journals are likely to decline a review of articles which have been reviewed and rated by another journal. If journals aim to market interesting articles to their readers then they would be able to increase the reader's satisfaction by informing them about stellar articles in their field, even in cases, where the journal was not the first reviewer. A top journal can increase the satisfaction of their readers by including top articles which have been published in a competitor's journal. As long as the articles are accessible to the public, no additional costs are incurred by including articles from a competitor's journal. This requires that a journal analyze the reliability of its competitor's review processes. If necessary, it may have to decide how to translate the review result of its competitor into its own review framework. Possibly, a standard review schema may emerge. In short, articles may be published in several journals because the readership community of several journals may profit from the same article. The consequence is **proposition #11**: *An article may appear simultaneously in several journals.*

The reengineered publishing process for online journals is shown in Figure 5:

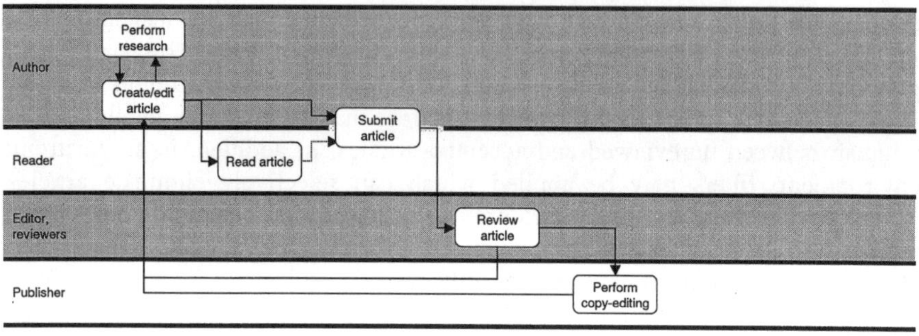

Figure 5. Publishing process for online journals

2.1.2.3. Organization perspective

2.1.2.3.1. Stakeholders

The main stakeholders of online journals can be identified by examining the reengineered process. Only four of the seven initial stakeholders are central to the online publishing process. The fate of publishers, distributors and libraries is uncertain. While they may not be core participants of the scientific communication process, it is likely that publishers and libraries will identify new niches. Among others, publishers may focus on "providing certain types of search tools"[and] "providing accessible permanent storage" (Braman 1995, p 47) or they may become a clearing-house for specialized information (Horowitz 1986, p 33).

The issues are different in the case of libraries which have a long history as repositories of knowledge. As the internet provides access to information from anywhere, the library is not likely to persist as a location. Even on the internet, however, much information is not accessible for free. Libraries may negotiate and administer collective access to fee-based resources or may become the managers of a university's intellectual property rights (Arnold 1993, p 78). Another traditional role of libraries has been the development and maintenance of classification systems and catalogs. Much progress has been made in the standardization of classification systems and the typical university library has ceased to devote significant resources to their extension and maintenance. Could the electronic environment lead to a rebirth of this role? Some factors are in favor: as discussed above, the new environment may shift the focus toward the synthesis of knowledge. Scientists need knowledge maps and effective classification systems to access knowledge available on the internet. While librarians might be ideal for this task they might not have enough knowledge in the subject areas for carrying out the synthesis or the development of knowledge structures. In the electronic environment, dynamic, multi-dimensional knowledge structures may emerge that are much more complicated than traditional classification systems. Ultimately, the only group capable of synthesis may be the researchers themselves. There may be room for libraries to develop general, high level knowledge structures which apply to many subject areas. In an electronic environment, libraries may see their role in the development of standards and access mechanisms for knowledge. Another alternative for the library of the future is to become an access mechanism to knowledge in a specific field. A library may choose, for example, to broker knowledge in a field in which the university is a recognized leader. The library would work with the faculty to develop and continuously update the knowledge structure for that field. This would increase the prestige and the scientific relationships of the university in this field. It would require cooperation with journals (electronic and traditional) and other sources of knowledge. It might lead to competition with professional associations.

2.1.2.3.2. Objectives

Information technology changes process and power differentials within the publishing system. Therefore the relative weight of objectives may change. The decline in the publisher's importance may lead to a reduction of economic objectives. The ability of the author to self-publish, may provide more weight to his objectives – such as being able to create new knowledge. On the other hand, quality control objectives may suffer. This may happen gradually – as authors discover new mechanisms to self-publish and to bypass established journals to reach their target audience directly – e.g. via search engines. While the research community needs to be concerned about such an evolution, it needs to be stressed that such a development does not necessarily reduce the usefulness of the scientific body of knowledge. It is questionable whether large efforts for quality control are efficient when a high percentage of articles are never cited anyway. A study conducted by Garfield, for example, found that 56% of the articles indexed by the Institute of Life Sciences between 1955 and 1987 were never cited – not even self-referenced (Garfield 1989). An earlier study arrived at the same conclusion (Menard 1971, p 21). Another impact of information technology may be an increased emphasis on the synthesis of knowledge.

The changes described above will lead to changes in the academic reward system. Current academic rewards are strongly tied to the number of articles in leading scientific journals. However, when variance in structure and size of contributions increases, then the number of publications ceases to be a good productivity indicator. As the primary roles of researchers become more diverse, it also becomes more difficult to compare outputs. While the current ideal are *original* contributions, in the future the synthesis of scientific knowledge may be recognized as an equally important contribution. Furthermore, in an electronic environment, reactions of readers and practitioners are more directly accessible. Evaluations can be multidimensional. Also, many additional metrics for the quality of a contribution such as the number of citations can be determined. The structural analysis of knowledge may yield more accurate information about how central a contribution is in the body of knowledge.

On the other hand, the current reward system may limit the adoption of new features. Online journals are currently less respected than established journals. Therefore authors forego some of their expected rewards when publishing in such journals (Harter & Park 2000). The higher the quality of an article, the larger the sacrifice; therefore online journals often have difficulties in attracting high-quality contributions. The move towards online knowledge infrastructures challenges the scientific community to develop new standards for rewards. In other industries, most notably in consulting, companies have already changed their reward structure to reflect the additional needs in synthesizing and communicating knowledge. In the scientific community, this change will most likely be gradual and take considerable time.

2.2. Empirical analysis: Emerging knowledge infrastructures

> "*The new electronic medium gives us the opportunity to reconsider many aspects of our current research communication, and researchers should take advantage of this opportunity to map out the ideal research communication medium of the future.*" ... "*In particular, we need to dislodge definitively the curiously prevalent notion that the future electronic medium will strictly duplicate, inadequacy for inadequacy, the current print medium.*" (Ginsparg 1996)

The theoretical analysis in the preceding section has shown the potential for significant change in the academic publishing system. The analysis led to the creation of a theoretical concept – a mature electronic journal – to which online journals may converge. Although the theory is clearly falsifiable by future developments, it needs to be complemented by an analysis of actual knowledge infrastructures which are currently emerging on the internet. On the internet, many sites are now available which aim to support the creation, synthesis or dissemination of scientific knowledge. While these infrastructures differ significantly in purpose, approach, technical architecture and functionality, they can be a rich source of insights about knowledge management in an IT-based environment.

Ideally, an analysis of a sample of infrastructures would lead to a taxonomy of scientific knowledge infrastructures. Characteristics of each current exemplar would be analyzed and clustering methods would be applied to identify subsets of knowledge infrastructures with similar characteristics. Unfortunately, such an approach faces three major obstacles:

The first limitation of such an analysis is the absence of equilibrium. In the current historical situation, the internet and related technologies advance rapidly. In

1996, for example, major effort was required to link a database to the World-Wide Web. Two years later, off-the shelf middleware was readily available to connect databases reliably to the web. Most web server software incorporates database connection functionality. This has significantly increased the spectrum of possible applications for knowledge infrastructures and many infrastructures are now built around databases. Technological innovation continues further. At present many additional advances – e.g. a maturing Java programming language, support for XML, new search technologies – further increase the space of possible solutions for knowledge infrastructures. Current infrastructures are not mature. Therefore they cannot be regarded as representative of future online infrastructures.

This problem is compounded by the time lag for exploiting technical advances. Solutions are not always immediately evident and their implementation consumes time. Sites may reflect technological constraints which were present when the sites were developed but which meanwhile have been removed. This limits the inferences that can be drawn from observation. A third problem is that knowledge infrastructures need to integrate themselves into current scientific traditions. Designers of knowledge infrastructures recognize that they are rarely able to invoke drastic change in established processes. To be successful, their solutions need to be compatible with existing scientific traditions. Thus, they may hold back conceptualizations of technically and organizationally feasible knowledge infrastructures, for which 'the time has not yet come'. They may replicate inefficient paper-centric solutions to match traditional processes. Thus some of the features of current online infrastructures may be bridges to the past rather than characteristics of a novel medium.

Other fields besides information systems have encountered the problem of unstable categories in rapidly evolving domains. In biology, many methods have been developed for creating taxonomies (Eldredge & Cracraft 1980). Only in the beginning did these methods focus on observable characteristics (phenotype). Later, the analysis shifted towards more fundamental phylo-genetic characteristics which are rooted in the theory of evolution. In biology, the genetic relationship between ancestor and descendant now forms the basis for taxonomic analysis.

Unfortunately, no comparable theory of evolution is available for IT-based infrastructures. In IT change is not limited by genetic relationships. Characteristics which are available in one knowledge infrastructure can be replicated easily and rapidly by other infrastructures. For example, most online journals initially did not allow articles to be changed. Now an increasing number of online journals have policies which outline under what conditions articles can be changed and some journals have introduced version management systems. As such policies become commonplace, authors or readers exert pressure on lagging journals to implement the corresponding mechanisms. Another example for rapid change is search functionality. Initially few online journals had a search engine. As search technology matures, most online journals incorporate increasingly sophisticated search capability. Therefore, in the absence of a recognized mechanism for the evolution of IT infrastructures, a path that is similar to the phylogenetic analysis in biology cannot be followed for the analysis of knowledge infrastructures.

Because of both issues – the rapid maturing of technology and the absence of a known evolutionary mechanism, the current set of exemplars of knowledge infrastructures cannot be treated as representative of the set of all such infrastructures. This significantly limits the inferences which can be drawn from a snapshot of current knowledge infrastructures.

As a consequence, it is premature to empirically identify categories of knowledge infrastructures. Nevertheless, an understanding of the spectrum of current infrastructures is important for those who currently develop knowledge infrastructures and those who track their evolution. It is also important for identifying potential flaws in the theory of online journals developed in the previous section and for refining this theory.

A more realistic goal of empirical analysis is to examine the range of characteristics of current knowledge infrastructures and the range of their possible configurations. By focusing on distinguishing characteristics it is both possible to spot isolated, highly innovative approaches which are the forebears of tomorrow's state of the art as well as to identify evolution paths among the knowledge bases that depend on factors other than technology.

To identify the spectrum of possible knowledge-related sites, characteristics of current sites will be identified in the following. Characteristics are elicited by comparing electronic sites with traditional media and by comparing different electronic sites. The characteristics will then be prioritized and organized into groups of key characteristics. This allows to extrapolate the full spectrum of possible knowledge related sites by considering all combinations of independent characteristics. In the following, sites will be examined by categories typically mentioned in the literature such as: electronic journal, digital library, knowledge repository, collaboratory. Then sites will be examined which do not easily fit into these categories. Before synthesizing the characteristics, visions of future knowledge infrastructures articulated by several researchers will be analyzed to determine potential characteristics which have not yet been implemented in current sites. Figure 6 shows the categories of infrastructures which will be evaluated.

2.2.1. Electronic journal

The internet abounds with sites that are dedicated to scientific knowledge. Many of these sites characterize themselves as 'electronic' (or 'online') journals. The term, which may have first been proposed by Sondak and Schwartz (1973), is used with several different connotations. Some authors employ the term for the vision of scholarly IT based infrastructures which will eventually replace paper journals (Watson 1994). In this sense, most of the different types of knowledge infrastructures which are discussed in this chapter could be characterized as electronic journals; therefore the more generic term 'knowledge infrastructure' will be used for this interpretation in the following sections. In most cases the term 'electronic journal' is used more narrowly for sites which replicate structures typical for traditional paper-based journals. Thus electronic journals are the electronic network-based analogue of traditional paper-based journals. They share the same goal of being a "primary channel for communication of knowledge in its field" [and becoming] "one of the arbitrators for the authenticity or legitimacy of that knowledge." (LaFollette 1992, p 69) Like paper journals, e-journals are collections of largely independent articles, are frequently organized into issues, publish in regular intervals and adopt some variant of peer review. While not replacing the model of traditional journals, many electronic journals leverage information technology to improve on its features. Innovative electronic journals, for example, implement electronic submission and accelerated review processes which may include the ability to perform micro-reviews (Barua, Chellappa & Whinston 1997).

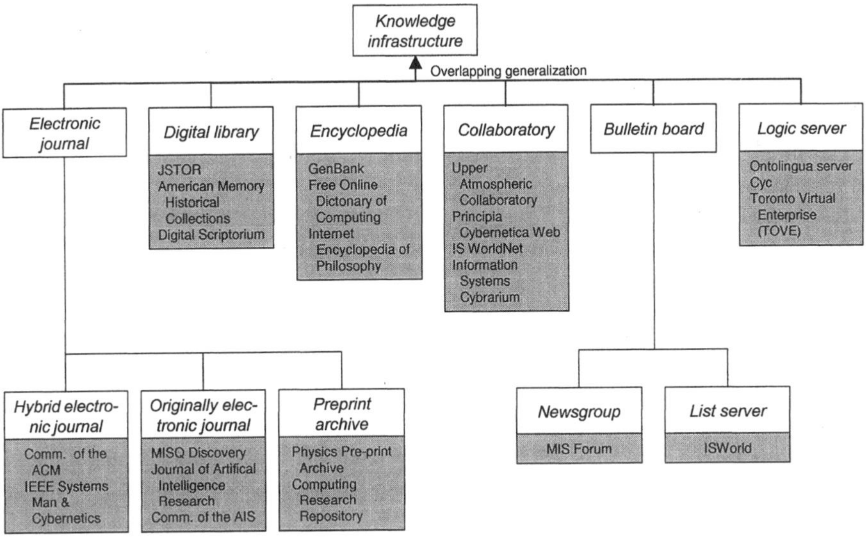

Figure 6. Categories of knowledge infrastructures

Many journals which are currently available in full text online are electronic versions of paper journals. These journals are 'hybrid' electronic journals because their online activities are determined in large part by established practices in publishing the paper version. In recent years, however, many new journals have been established online which are not dependent on a paper version. Such journals will be called 'originally' electronic journals.

2.2.1.1. Hybrid electronic journals

Most publishers of paper journals have begun making online versions of their paper journals available. Examples are the Association for Computing Machinery (CACM), IEEE, Elsevier and Springer (see Table 6). These sites are hybrid sites; journals are published both on the internet and in paper. In most cases access is restricted to subscribers of the electronic journal (or an electronic library). ACM also provides pay-per-view access to individual articles. The presentation of the journals on the web resembles the traditional structure. They are organized by volume and issues: tables of contents provide access to individual articles. In most cases, access to abstracts and keywords is free. Articles typically have an index page which shows title, abstract and keywords and provides a link to the full text. The latter is mostly available in PDF or Postscript format but less frequently in HTML. As an alternative form of accessing articles, most sites provide a search engine. Search technology is advancing rapidly and the range of capabilities varies considerably. Most search engines provide full-text search; some provide context-based search or searches by article attributes such as publishing date, keywords etc.

Table 6. Number of hybrid electronic journals available by major publishers

Publisher	# journals	Source
ACM	30	http://www.acm.org/dl/toc.html
Cambridge Univ Press	100	http://journals.cambridge.org
Elsevier	1816	http://www.sciencedirect.com
Taylor & Francis	740	http://www.tandf.co.uk/journals/default.html
IEEE	132	http://www.ieee.org/products/onlinepubs/iel/iel_ieee_pubs.html
John Hopkins Univ. Press	217	http://muse.jhu.edu/journals
Emerald (MCB Press)	108	http://konstanza.emeraldinsight.com
Oxford University Press	178	http://www3.oup.co.uk/jnls/online
Springer-Verlag	513	http://link.springer.de/ol/total/print.shtm
Taylor & Francis	740	http://www.tandf.co.uk/journals/default.html
Wiley Interscience	365	http://www3.interscience.wiley.com/cgi-bin/browsepj
Sum	4199	

List of publishers extended from: Hitchcock, Carr & Hall (1997). Numbers of journals determined from the tables of contents at the publishers web sites on 12/16/2002.

Hybrid sites are necessarily very close to paper journals. Most elements of the publishing process remain unchanged. In the case of the above mentioned journals, for example, accepted papers cannot be changed. The review process remains essentially the same (peer review) although some of the journals now accept online submissions and some of the sites are moving toward online support for the review process.

2.2.1.2. Originally electronic journal

An increasing number of online journals have been established without being based on a pre-existing paper journal. Such journals will be called 'originally electronic journals'.

2.2.1.2.1. Journal of Artificial Intelligence Research

The Journal of Artificial Intelligence Research (JAIR, www.jair.org) is one of the early online journals which were established without being based on an existing paper journal. Content is available in PostScript, PDF and occasionally in HTML. The full content is provided at no cost to the reader. Similar to traditional journals, the table of contents is organized by volume and author. However, the journal has abandoned the issues metaphor. Recently an innovative subject area index has been added: readers can browse a JAVA applet which graphically clusters articles by subject area. A screen shot of this 'information space' is shown in **Figure 7**. Articles are represented in clusters around topics such as 'Problem solving', 'Deduction and theorem proving', 'Natural language processing'. When readers click on individual items in a cluster the title and abstract of the associated article are displayed. In JAIR, articles are submitted electronically – either as email attachment or as URL pointing to the manuscript. A traditional peer review process follows. Once papers are accepted, they are published individually without further delay. As a bridge to the current paper-centric environment, JAIR publishes a paper version of the latest volume at the end of each year. For more detailed information on JAIR see Kling & McKim (2000) and Kling & Covi (1995).

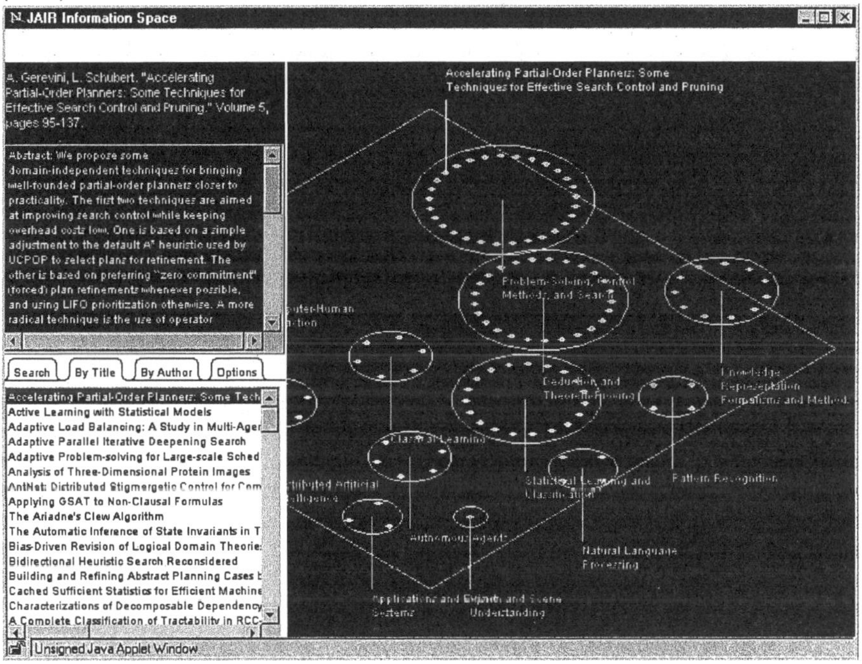

Figure 7. JAIR Information space (www.infoarch.ai.mit.edu/jair/jair-space.html, 5/7/2002).

2.2.1.2.2. MISQ Discovery

MISQ Discovery (www.misq.org/discovery) is a venture of MISQ, one of the leading information systems research journals. It was announced in 1994 as "an adventuresome and experimental electronic production" (Ives 1994 p xxxviii) which, among others was intended to "challenge the foundations of paper-based scholarship." (Ives 1994 p xxxviii) The first article was published in September 1996 (Ytterstad et al. 1996). By September of 1998, only one more article had been published. The journal, however, has introduced important innovations: It pioneered the distinction of living versus archival articles. Archival articles can only be subsequently edited to correct typing, spelling, grammar, or to update links (www.misq.org/discovery/about.html#living, 9/17/2002). In contrast, living publications are kept up-to date by the authors. In addition, for the first article a listserver-based bulletin board was provided for the discussion of the paper. This feature was discontinued, however, because of apparent lack of interest. MISQ Discovery uses a traditional peer-review system. To improve the acceptance of the journal, an abstract of each accepted article is published in MISQ and the article becomes part of the MISQ index.

2.2.1.3. Pre-print archive

Another variant of web-based academic journals are pre-print archives, sometimes also called e-print servers (Buckingham Shum, Motta & Domingue 2000). They are used for the electronic distribution of articles among scientists before they are submitted to a journal for 'official' publication. At present several such archives exist. The first and most successful pre-print archives was started in 1991 by Ginsparg, a physicist at Los Alamos National Laboratory. The "Physics pre print archive" is accessible at xxx.lanl.gov and was initially limited to pre-prints in High Energy Theoretical Physics but since has been expanded significantly. Ginsparg describes the situation in which he started the archive as follows:

> "At least since the mid-1970s, the primary means of communicating new research ideas and results had been a pre-print distribution system in which printed copies of papers were sent via ordinary mail to large distribution lists at the same time that they were submitted to journals for publication." (Ginsparg 1994, p 390)

Ginsparg's innovation was a list server which archived and distributed pre-prints in PostScript or Tex formats. Thus this electronic knowledge infrastructure was an extension of the traditional preprint journals which emerged in the Seventies in physics (Harnad 1995). It greatly simplified the distribution of pre-prints. In contrast to conventional discussion forums, readers were able to formulate queries and to retrieve back issues. Subsequently, a web-based interface was added which further simplified access to all papers in the archive.

The archive is structured into more then 30 subject areas. It recently has added sections for other sciences (e.g. mathematics, economics and recently computer science (Halpern 1998)). Authors submit their finished articles to one of the subject areas. Mechanisms for cross-listing a pre-print in several sections exists. Readers are presented with daily lists of new submissions which show title, authors and abstract for each pre-print and which contain a link to the text file. The system has search capabilities. The archive supports multiple versions of a document. It is possible to change preprints after submission; new versions receive a new identifier. However, pre-prints cannot be directly linked to one another. To discourage authors from distributing low-quality articles, authors cannot delete articles once they have been submitted.

The archive has been extremely successful. In the last quarter of 2002, the average number of weekly connections exceeded 600.000 (arxiv.org/show_weekly_graph, 12/14/2002). In December of 2002, the archive was mirrored in 17 countries. Ginsparg reports that one result of the archive is that "many institutions have already eliminated their hard-copy distribution of pre-prints." (Ginsparg 1994, p 394)

This has led to changes for academic journals in physics: Several scientific physics journals have started to accept preprints from the archive for submission. To submit a paper to one of the American Physical Society Journals (publish.aps.org/esubs/guidelines.html, 4/26/2002) or to the Journal of High Energy Physics (jhep.cern.ch/JOURNAL/xxx.html, 4/26/2002)., for example, an author can electronically submit the identifier of his pre-print. The editor then downloads the document from the archive.

In the mean time, many other pre-print archives have been founded. An 'open archive initiave' has been started with the goal of integrating pre-print archives and to allow harvesting of meta-data (Van de Sompel & Lagoze 2001). With the success of pre-print archives, Ginsparg has become a fervent advocate of redesigning the intellectual infrastructure in science. He is convinced of the "irrelevance of refereed journals to ongoing research" (Ginsparg 1994, p 390) and argues that on the internet, information is practically free. "The entire journal structure and organization needs to be reconsidered in light of the electronic format." (Ginsparg 1994)

2.2.2. Digital library

"The objective of an electronic library is to bring the library to the user" (Story et al. 1992, p 17). The main functions of a digital library are providing access to large collections of documents and to organize them. This may involve many challenges, including the digitization of print documents, providing uniform access to documents in very different electronic formats, negotiating terms for accessing distributed collections from various publishers and providing search functionality and services.

Several projects are currently underway to provide access to large collections of documents (Treloar 1998). In the US, the National Digital Library Program at the Library of Congress has created a collection of more than 100 American Memory Historical Collections which provide free access to historical documents, photographs and movies (memory.loc.gov/ammem/ammemhome.html, 12/11/2002). The Digital Scriptorium is a collaboration between the libraries at UC Berkely and University of Columbia to provide access to digitized versions of their medieval and early renaissance manuscripts via the web. In December of 2002, more than 9000 digitized images and more than 2500 documents were available in the library (sunsite.berkeley.edu/scriptorium/form.html, 12/11/2002).

2.2.2.1. JSTOR

JSTOR is an archive of digitized (bitmapped) articles from print journals. The archive is limited to articles which are more than three years old. More recent articles have been excluded to minimize effect on the revenue stream of publishers while improving access to older articles. JSTOR was founded in 1995 with initial funding from the Mellon foundation. In 2002, JSTOR listed more than 1500 participant institutions and provide access to the content of more than 300 journals in 15 disciplines.

2.2.3. *Encyclopedia*

The traditional, paper-based environment, has developed an attractive alternative to the dissemination of knowledge via journals: the encyclopedia. Encyclopedias are "works that claim to provide in an orderly arrangement the essence of 'all that is known' on a subject or a group of subjects" (Brittannica Online 1998, www.eb.com:180/cgi-bin/g?DocF =macro/5002/11/0.html, 9/14/1998). By focusing on the organization of knowledge, encyclopedias follow an approach which is fundamentally different from the journal model. The first encyclopedias appeared after 1750 - about a hundred years after the first journals were founded. Diderot's 'Encyclopedie' significantly simplified access to knowledge by introducing cross-references – the precursor to modern hypertext (Guedon 1996, p 77).

Encyclopedias have become an important mechanism for storing knowledge in large and complex domains. They typically strive for complete coverage of a domain. Ideally, they provide information about every relevant phenomenon and concept. While most encyclopedias are organized alphabetically in dictionary form, other variants exist which are structured by topic. As most encyclopedias are organized alphabetically, they do not exhibit deep structure. This reduces the amount of effort needed to access information about any part but also makes the systematic acquisition of knowledge - learning - from an encyclopedia difficult. A problem of encyclopedias is that they exclude preferred sequences of reading. Encyclopedias and dictionaries maximize the number of entry points for the reader. The writer cannot assume that somebody who is reading a specific entry has read any other of the entries.

The two major differences in journals are the goal of presenting the knowledge of a domain in an organized, systematic fashion and the extensive cross-referencing between entries. Another difference between encyclopedias and journals is the way in which authorship is treated. While journals always disclose the authors of their articles, in an encyclopedia, authors often remain anonymous (or appear in a summary list for the whole encyclopedia). Although exceptions exist, this practice points to a different understanding of the role of authors. In an encyclopedia, authors are often seen as collectors and organizers of knowledge who constantly monitor the field in search for new information or information that has been missed in the previous edition. This is in contrast to the authors of scientific journal articles who often see their role as being the originators of 'new' knowledge. While one may dispute the idea that the synthesis of knowledge is not a creative process and does not lead to new knowledge, the more fundamental problem in encyclopedias is attributing authorship: When many persons collaborate on a contribution over years or centuries, then who is to be regarded as the author? The issue is not how to determine who wrote which words. In an electronic environment, it is technically possible to track authorship for each letter, word, sentence or figure. The problem is how to determine authorship for collections of thoughts, integrated bodies of knowledge to which many people have contributed some insights. Attributing authorship in a body of tightly integrated and continuously updated knowledge is much more problematic than in a collection of loosely organized, largely self-contained and static articles. Encyclopedias show that the concept of authorship may have to be rethought when science moves towards more integrated bodies of knowledge.

Another difference between journals and encyclopedias is the way in which they make use of information technology to organize knowledge and ensure consistency between entries. For example, algorithms may check whether cross references are complete so that for every item in one direction a corresponding reference exists in the other direction. Other algorithms may ensure that entries follow appropriate patterns or spot circular definitions (e.g. the term 'customer' is explained using the term 'client' and the term 'client' is explained with the term 'customer'). Information technology can spot such inconsistencies in many cases.

On the World Wide Web, several scientific sites have emerged which focus on the integration and organization of the knowledge for a limited domain. Three examples will be discussed subsequently. Extrapolating from the characteristics, the encyclopedia metaphor can be extended to yield a vision of future knowledge infrastructures:

> "A kind of living encyclopedia would progressively come to the fore, one where distinctions between teaching and research, as well as distinctions between domains and disciplines, would probably be deeply redefined. The limits of the encyclopedia would then become the shifting, moving limits of knowledge itself, and all involved with knowledge, be it teaching, learning or researching, would envision their work as intellectual moves within the abstract, multidimensional space corresponding to a humanity-wide hypertext. Publishing would lose its function of re-presentation to become an integral, immediate dimension of the dynamics of human knowledge at large." (Guedon 1996, p 83)

2.2.3.1. Free Online Dictionary of Computing

One example of an internet encyclopedia is the Free Online Dictionary of Computing (Foldoc, wombat.doc.ic.ac.uk). It has been available on the World-Wide Web since 1994 and contains more than ten thousand short but intensively cross-referenced entries. The dictionary has a single editor who maintains tight control of its content; contributions of other authors are solicited and a mechanism for guest editors has been set up, but entries do not list the author. A main page, however, lists more than 800 individual contributors. An innovative feature of the dictionary is a list of missing entries: when a reader requests a non-existing entry in a search form, the system records the search words and adds it to a list of missing entries. Contributors are encouraged to add descriptions for those terms entries. Contributors are encouraged to add descriptions for those terms with the highest frequency.

2.2.3.2. Internet Encyclopedia of Philosophy (IEP)

This encyclopedia of philosophy is an internet resource accessible at http://www.utm.edu/research/iep. Its entries are alphabetically structured. A timeline provides an alternative access path. In addition, a keyword index identifies related entries for each keyword. Although timeline and keyword index use hyperlinks, the entries themselves are not cross-referenced via hyperlinks. IEP invites submission from researchers to extend, replace or add new keywords. There are no size limitations and the authors retain the copyright. Access to IEP is free; readers are explicitly allowed to copy and distribute its materials for personal and classroom use. IEP provides full text search capability. A similar approach is taken by the Stanford Encyclopedia of Philosophy (plato.stanford.edu). Each entry is assigned to an author who has direct electronic access and is responsible of keeping it up to date. Editors are automatically notified about changes and additions. The result is a dynamic encyclopedia which is "responsive to new research" (plato.stanford.edu/contents.html 9/17/2002). While the Stanford Encyclopedia does not contain timeline or a separate keyword index, it contains a list of recent contributions organized by submission date.

2.2.3.3. GenBank

GenBank (www.ncbi.nlm.nih.gov/Web/Genbank) is an electronic database maintained by the National Center for Biotechnology Information in the United States. It contains more than one billion annotated and cross-referenced DNA sequences. Data is regularly exchanged with other DNA databases located in Europe and Japan. It was founded in 1970 by the physicist Walter Goad as a database of nucleic acid sequences. Later it was renamed the Los Alamos Sequence Library (LAS) and officially became GenBank in 1982 (National Research Council 1993, p 45).

GenBank pursues the following three goals

> "(1) an archive, for long-term storage and preservation of sequence data, (2) retrieval of nucleotide sequences, with delineation by sequence-specific, bibliographic, physical, or functional criteria; and (3) a research platform, for examination and analysis of sequence data." (Burks et al. 1990, p 4)

Initially GenBank operated as an abstraction service which searched the literature and entered sequence information into the database. This soon proved too time-consuming and in the early Eighties, GenBank switched to direct submission of sequences by researchers (Hilgartner 1995), pp 251-254). Scientists can submit their DNA sequences to GenBank free of charge on the web. Before a submitted sequence is published, it is reviewed and indexed by GenBank staff. Not all GenBank submissions are accessible by the public. On demand by the author, access to a sequence can be blocked (www.ncbi.nlm.nih.gov/Web/Newsltr/feb98.html #GenBank, 6/23/2002). GenBank incorporates a revision mechanisms to allow authors to update their sequences. The success and importance of GenBank has led many scientific journals to require a submission to GenBank as a prerequisite for publishing an article. This secures access to the data to the scientific community while at the same time reducing the need to publish this data together with the printed article (Cummings et al. 1992, pp 123f).

An important functionality for gene researchers is the ability to perform similarity searches between DNA and protein sequences using software which was specifically developed for this purpose. The National Research Council used the example of GenBank to illustrate the potential of information technology:

> "The most important shift to electronic data publishing is recruiting of the community of researchers to perform the job of sequence entry and annotation and database building." ... "The scientists who have determined the sequences are also the most appropriate people to annotate their own work, given a controlled semantics and syntax." (National Research Council 1993, p 47)

While GenBank was not intended as a competitor to academic journals, it has improved the ability of the research community for gathering high-quality data. Although no systematic studies have been executed yet, it may have shifted the focus of research efforts more towards data. Cinkosky et al. concluded: "Our experience indicates that electronic data publishing has actually resulted in a higher level of quality for data in GenBank and therefore also in journals." (Cinkosky et al. 1991)

2.2.4. Collaboratory

Collaboratories are often regarded as the most advanced form of knowledge infrastructures. The term was coined by an assistant director at the National Science Foundation, William Wulf, in a white paper on a 'national collaboratory' (National Research Council 1993), p vii). The word is a combination of the terms 'collaboration' and 'laboratory' and represents

> "a center without walls, in which the nation's researchers can perform their research without regard to geographical locations - interacting with colleagues, accessing instrumentation, sharing data and computational resources [and] accessing information in digital libraries." (Wulf 1989; see also Ross-Flanigan 1998)

The concept was the subject of several NSF workshops (Lederberg & Uncapher 1989; Finholt & Olson 1997) and led to three implementation projects, one of which – the Upper Atmospheric Collaboratory (UARC) - will be described below.

In contrast to electronic journals and encyclopedias which are focused primarily on the dissemination of knowledge, collaboratories directly support the research process, the creation of knowledge and the communication between researchers. Barua et al emphasize this collaborative function when they describe a collaboratory as "an open electronic platform for individuals or groups with common interests to efficiently exchange, disseminate and create issues, ideas and knowledge." (Barua, Chellappa & Whinston 1995, p 418) In collaboratories, knowledge ceases to be static and final. Information relating to research programs, unresolved problems, data and preliminary results are equally important. This significantly changes the nature of such knowledge infrastructures; they become highly interactive and much more fluid in their content.

Although several internet sites describe themselves as 'collaboratories' the term is often limited to communication between researchers. Only in this sense CREC's MIS Forum which supports communication among information systems researchers and BioMOO (bioinformatics.weizmann.ac.il/BioMOO) at the Weizmann Institute which allows online discussion about Biology are collaboratories.

2.2.4.1. Space Physics & Aeronomics Collaboratory

After the need for collaboratories had been identified at the end of the eighties (National Research Council 1993), space physics was one of the first areas where the concept was implemented. With funds from the National Science Foundation, the Upper Atmospherice Collaboratory (UARC later renamed to Space Physics & Aeronomics Collaboratory SPARC) was started in 1992 at the University of Michigan in collaboration with several other research institutes around the world. The first step consisted in tying together instruments operated in various sites and allowing uniform access to the stream of data which they generate. The infrastructure subsequently was extended to include chat capabilities and Java-based applets for the visualization of data. In addition, access to computer models was provided. Researchers typically work together in tightly coordinated research campaigns which last for days or weeks. Research objectives are defined beforehand and access to instruments is coordinated. During the campaign, researchers are able to communicate via chat windows to interpret data on the spot. An example of a screen used during a 1997 campaign which purposefully tied together instruments for specific research objectives is shown in Figure 8. The top left window contains the session manager. Below is a chat window. In addition, satellite data and a TNG model are shown on top. On the lower part, several data from data from scatter radar at three different locations is visible.

Communication between collaborators and usage of SPARC has been intensively studied. Key advantages are the ability to leverage expertise from researchers who are distributed across the globe and the ability to expose graduate students much earlier to empirical data and the problems of data collection. Campaigns which require intensive, coordinated research in short periods may significantly increase research productivity. Collaboratories are not without problems, however. Electronic communication alone fails to build needed trust and personal relationships between participants (Ross-Flanigan 1998).

2.2.4.2. Principia Cybernetica Web

Principia Cybernetica Web is one of the early web-based knowledge infrastructures. It was established in 1993 aiming at "the computer-supported collaborative development of an evolutionary-systemic philosophy." (pespmc1.vub.ac.be, 12/27/2002) It now consists of more than 2000 mostly hierarchically organized nodes addressing topics in cybernetics and philosophy. The system allows any reader to add comments to any node (including comment nodes) without review. More refined contributions are published after review. However, as one of the editors points out, the rejection rate can be low compared to traditional journals:

> "Although the editors have the right to reject material they deem inappropriate or of insufficient quality, the low cost of electronic publication allows us to be quite liberal. As long as there is sufficient storage space, all contributions can be published." (Heylighen & Bollen 1995)

As the list of recent changes (pespmc1.vub.ac.be/recent.html,12/27/2002) shows, the site continues to be expanded regularly. Unfortunately, current site statistics are not available. The site contains several innovative features. A 'hit parade' lists the most popular nodes. The system also contains a random navigation feature which returns a random node to the reader who is browsing without a particular goal. An imagemap graphically shows the top level of the node hierarchy and allows direct access to these nodes.

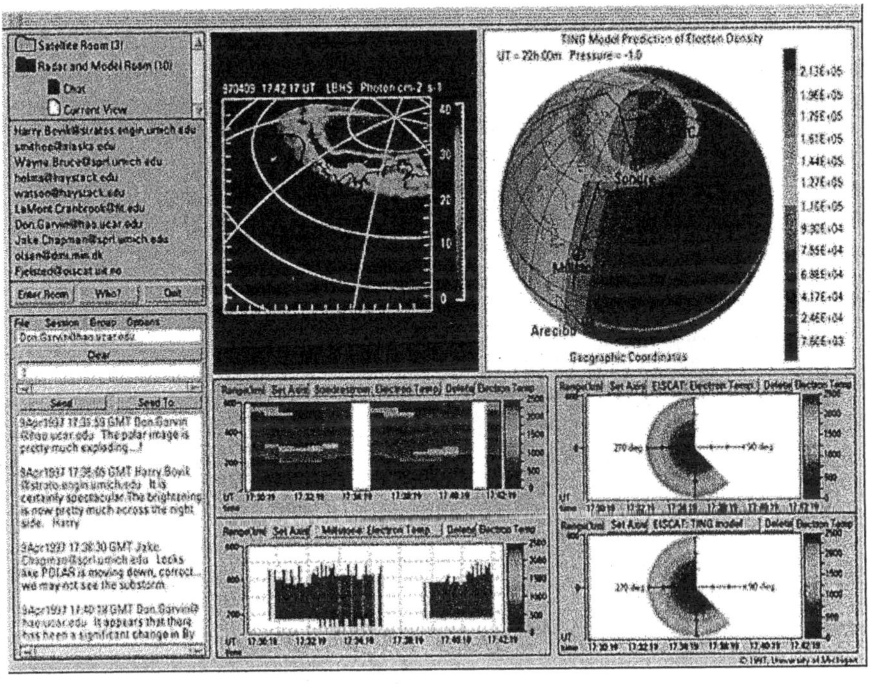

Figure 8. Screenshot of a UARC campaign interface (geneva.crew.umich.edu/UARC/background.htm 8/7/1998)

In addition, Principia Cybernetica operates two mailing lists. It has acquired the copyright to several out of print books and republished them on the web. An example is DeRosnay (1979). Its full text is now available at: pespmc1.vub.ac.be/MACRBOOK.html (12/27/2002). A compressed version of the whole site is available for downloading and local installation. Moreover, a printable version is in the works. In summary, Principia Cybernatica is a highly innovative knowledge infrastructure which increases the level of interaction through the ability for direct annotation and joint work on a common topic. The degree of interaction, however, is significantly lower than in UARC.

2.2.4.3. Information Systems Cybrarium

This collaboratory is a knowledge infrastructure for the creation, synthesis and application of scientific knowledge in the field of information systems which has been developed by the author in the context of the current research. It allows a high levels of interactions between researchers and emphasizes the colloborative organization of knowledge. The system will be described in detail in section 4.1. below.

2.2.4.4. ISWorldNet

In 1994 the editors of MIS Quarterly focused particular attention on the impact of information technology on the scholarly infrastructure. Watson (1994) challenged the IS community to use information technology to create a global community of scholars. In the same issue, Ives (1994) outlined a plan to develop an academic infrastructure targeted at the IS community which he called "MISQ central". In the wake of the emergence of the world-wide web, Ives secured support from five academic information systems organizations to found a web-based infrastructure called ISWorldNet. Its mission was described as:

> "(1) provide information management scholars and practitioners with a single entry point to resources related to information systems technology, practice and knowledge; and
>
> (2) promote the development of a global platform to dramatically improve the world's ability to use information technology in creating, disseminating and applying knowledge." (Ives & Zmud 1994, p liv)

The founders hoped for quick adoption of the infrastructure by the IS community. They argued that the academic reward system would need to be changed to reflect ISWorldNet related efforts and themselves put substantial resources into its development in the next years. While ISWorldNet has been a considerable success in mindshare in the IS community, it did not grow as quickly as anticipated. In 1997, ISWorldNet server statistics were discontinued. Ives remarked that ISWorldNet has been disappointing in its implementation, citing lack of rewards, lack of focus and lack of structural depth (Chen, Ives & Leleux 1997). In April 1998 the ISWorldNet list of volunteers was smaller than 100.

ISWorldNet is hierarchically organized. It consists of three divisions (research and scholarship, teaching and learning, professional activities). These divisions are headed by a division editor and further broken down in sections and subsections with their own editors. Authors are encouraged to submit "bids" for certain themes in the form of html pages. When a bid is received, links to the bid are created. After a review from two editors (e.g. section and division editor), possibly one additional volunteer and taking into account comments from within the IS community, the bid is either accepted or rejected. An acceptance is not permanent; it may be canceled in the future.

Technically, ISWorldNet follows the standard WWW hypertext model. It consists of HTML pages with hyperlinks. Currently it only contains few dynamic or database components. There are detailed guidelines for the design of the HTML pages. Together with its hierarchical structure these factors reduce the degree of interactivity and involvement.

ISWorldnet continues to introduce innovations, however. It has introduced an electronic virtual meeting center where regularly weekly hour-long discussions are held about specific topics. It is a repository of valuable information concerning professional events, email addresses etc. Links to working papers can be directly submitted to ISWorldnet without review. In addition, local versions of ISWorldNet have been created in many countries.

2.2.5. Bulletin board

Bulletin boards and their subcategories newsgroups and list servers refer to technically and organizationally clearly defined services which can be found in very large numbers on the internet. All three services emphasize communication over the organization and storage of knowledge. Bulletin boards are systems which allow users to submit documents to an electronic white board (usually shown in list format) which then can be viewed by other users. News group are special types of bulletin boards which allow the annotation of documents in the form of threaded discussions. While newsgroups use the 'pull principle' for retrieving information, list servers electronicallly distribute new documents to the reader. Bulletin boards have become a widely used form of communication. Although many bulletin boards maintain archives of previous discussions and although many internet search engines include newsgroups in their index, contributions are not organized; as a consequence bulletin boards cannot be effectively used as a knowledge repository. The frequency of access to a document therefore decays rapidly with time.

Bulletin boards have been very successful. This is not so much an indication that users are willing to spend significant effort for finding answers to questions even if they are hidden among many contributions which are not relevant for their problem. Rather their success points to the critical role which communities and communications play in the utilization of knowledge. Bulletin boards establish a web of relationships among the active participants. They extend the horizons of novices and experts who monitor the board by identifying issues which they have not been aware of. For many persons, newsgroups have become an indispensable tool for continued learning. Bulletin boards show that in a knowledge infrastructure the importance of communication between members of the community must not be neglected but rather combined with knowledge repository functionality.

2.2.6. Logic server

Knowledge infrastructures differ in their approach for representing knowledge. While most scientific infrastructures focus on text and articles, some infrastructures adopt logic-based representation mechanisms. Because of the lack of an established term such systems will be called 'logic servers'. While such systems are few and – with the exception of Ontolingua Server (see below) – are limited to small groups of researchers, they exhibit characteristics which are very different from the other types of knowledge infrastructures. Logic servers attempt to realize the ideal of an integrated and largely consistent body of knowledge. Thus they impose strict formalisms for the description of knowledge and provide mechanisms for identifying dependencies and inconsistencies between statements. A disadvantage is the significant increase in effort for formulating knowledge. In addition, logic is less adequate for the social sciences (see section 3.2.2. beginning on page 143 for a more detailed review of the problems). Nevertheless, logic servers point to structures and functionalities which may be useful for other online scientific knowledge infrastructures.

2.2.6.1. Ontolingua server

This internet-based knowledge infrastructure has been developed by Stanford's Knowledge Systems Lab (KSL) and has been online since 1995 (Farquhar, Fikes & Rice 1996). Ontolingua server is a system for the collaborative creation, maintenance and application of formal ontologies. An ontology is a conceptualization of a subject domain; it is necessary, for example, for the development of agent-based systems. As ontology development requires significant effort and because considerable overlap between different domains can be expected, systems are needed which facilitate the creation and sharing of ontologies. Ontolingua server is based on a refinement of Ontolingua, a language developed by Gruber (Gruber 1993) for the specification of Ontologies, which in turn was based on KIF (Knowledge Interchange Format).

Ontolingua Server is an internet-based client-server system. Users can access an ontology editor via a standard web browser. Users must be registered to create their own ontologies or to view existing public ontologies. During ontology creation it is possible to incorporate other public ontologies, such as a Simple Time Ontology or an ontology which incorporates the internationally standardized physical units of the SI system. A screen shot of the Ontolingua editor is shown in Figure 9. A graphical menu is on the top of the screen. The lower part shows a summary of a Product Ontology. It lists several other ontologies which are used in the Product ontology. Barely visible at the bottom of the screen is the next heading which lists the hierarchy of classes belonging to the ontology. Ontologies can be manipulated in parallel by several members of a group. Participants are automatically notified of changes by other group members. Ontolingua server provides additional functionality for ontology analysis. For example, it is possible to compare different versions of the same ontology. Several features simplify the application of ontologies. Mechanisms are provided to translate ontologies into structures for typical application environments, for example, Prolog, CORBA, Clips, KIF etc. (Farquhar, Fikes & Rice 1996). In addition, applications can directly access the ontology via a published application programming interface (API), thus allowing remote ontology applications.

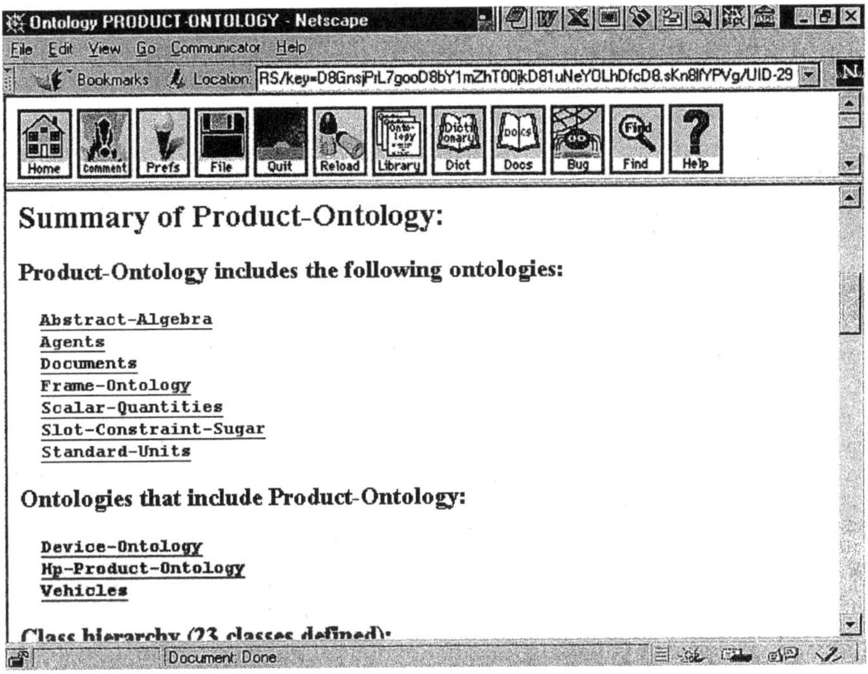

Figure 9. Ontolingua server example (http://www-ksl-svc.stanford.edu:5915, 11/1/97)

Ontolingua server is a unique knowledge infrastructure because of its support for the formal representation and integration of knowledge. It lacks mechanisms for review and critique of ontologies. A disadvantage of Ontolingua Server is its strong formal requirements which are not well suited for the capture of research findings in less formal disciplines such as the social sciences.

2.2.6.2. CYC

This is an example for a knowledge infrastructure that is intended to be able to reason independently and to provide answers to simple questions. From 1984 to 1994, Doug Lenat led a monumental effort of more than 100 person years to develop a knowledge base that would contain a significant amount of common sense knowledge. In that period, more than one million axioms have been entered into a custom-developed knowledge base (Lenat 1995). It was expected that at the end of this period the system would be cognizant enough to acquire additional knowledge through reading. While this point has not yet been reached significant progress toward this goal has been made and the CYC project has become a commercial enterprise (CYCORP).

In CYC, knowledge is organized in a "sea of axioms" formulated in a CycL, a language which is based on an extension of First Order Predicate Calculus. The knowledge base does not rely on frames. It is divided into a large number of contexts or "microtheories". This approach allows the inclusion of contradictory data and simplifies reasoning in limited domains. In addition, CYC contains a custom-developed inferencing engine and database as well as natural language interfaces (CYCORP 1999).

While the CYC knowledge base is proprietary, its core elements, the "Upper Cyc Ontology" have been published (www.cyc.com/cyc-2-1/toc.html, 5/5/2002). The Upper Ontology consists of about 3000 concepts which have been excerpted from the whole CYC knowledge base. They are structured into 43 groups such as 'Doing', 'Parts of Objects', 'Propositional attitudes' etc. The Upper Ontology is also accessible on Ontolingua Server.

A key problem with CYC is its ad-hoc approach in the development of the knowledge base. There does not seem to be an underlying principle in structuring the knowledge base and the creators have ignored much of previous research in ontologies (Sowa 1995). A significant percentage of the axioms were entered as random specific knowledge (Pratt 1994). Nevertheless, CYC indicates an additional path for the evolution of knowledge infrastructures. Knowledge may be formulated in the form axioms which then can be processed by reasoning algorithms. Extracting knowledge from the knowledge base then becomes a problem of inference rather then of retrieving the right set of matching documents. This has many implications; one of the main problems which it leads to is authorship. It may become irrelevant if knowledge is custom-assembled for each reader from a large number of axioms, and thus involves hundreds or thousands of authors.

2.2.6.3. Toronto Virtual Enterprise (TOVE)

Since the beginning of the Nineties, research at the Enterprise Integration Laboratory (EIL) in Toronto has focused on steps towards the development of a business infrastructure that enables the "dissemination of information, coordination of decisions, and management of actions to and among people within the organization and outside of it." (Enterprise Engineering Laboratory 1998) One core element for the development of such an infrastructure is the establishment of a common conceptual basis, an ontology, that can be used as the basis for communication between agents and actors in distributed enterprise environments. The development of a "shareable representation of knowledge" … "that minimizes ambiguity and maximizes understanding and precision in communication" (www.ie.utoronto.ca /EIL/comsen.html, 5/5/1998) is the goal of the ongoing Toronto Virtual Enterprise (TOVE) project.

An overview about the ontologies which are currently being developed is shown in Figure 10. All ontologies are based on a set of core ontologie which formalize key elements applicable to all domains of an enterprise. This level contains systematic representations of activities, resources, organizational concepts etc. These ontologies are subsequently combined and extended to create ontologies for specific application areas within an enterprise (e.g. inventory ontology) or for cross-enterprise applications such as an Enterprise Design Ontology. All ontologies are based on first order logic. Their axioms have been specified using Prolog. The process of ontology development has not yet been completed. In Figure 10, ontologies which have not yet been developed are marked with a "0".

Each ontology consists of several building blocks. The activity ontology, for example, is based on three foundational theories: Basic situational calculus, extended situational calculus and a theory of complex action. An excerpt of the situation calculus is shown in Figure 11. The first, axiom, for example, states that the result of any action a can never be the initial situation S_0. 'a?' represents an action variable, 's?' represents a situation variable. The do() function returns the event that results from performing a certain action in a starting situation s. This axiom implies that every action must change the situation. Activities that preserve situations (e.g. protecting, storing, etc.) cannot be modeled as actions. The second axiom defines that if two actions result in the same situation, then the actions must be the same, independently of the starting situations.

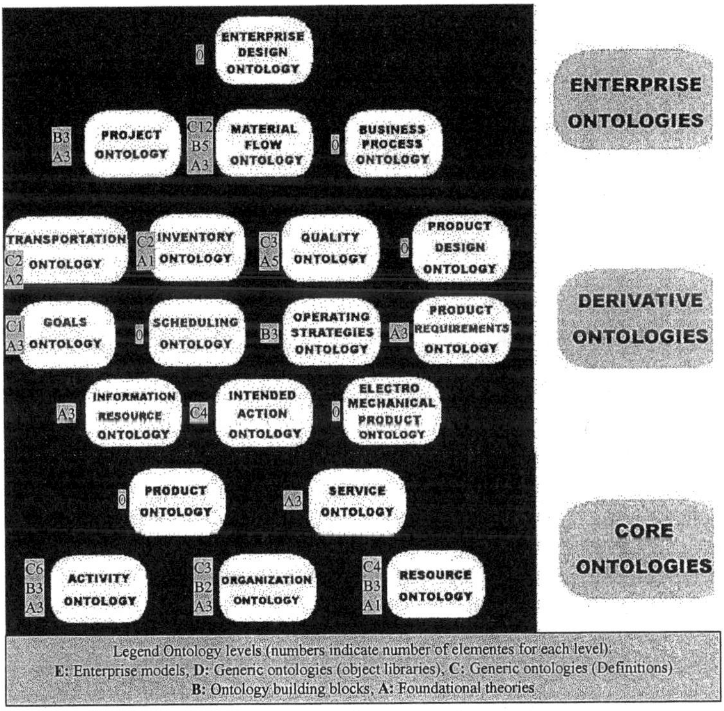

Figure 10. Elements of the TOVE ontology (adapted from http://www.ie.utoronto.ca/EIL/tove/toveont.html (5/5/1998))

Ontology building blocks then extend the foundational theories. The activity ontology, for example contains axioms which enable the specification of state formulae, complex action theories and occurrence theories. On the next level, generic ontology definitions follow. For the activity ontology, 6 sets of such definitions are available. The first, for example, concerns "Classes of ordered actions: Partially ordered actions". It contains definitions for the classes, functions and relationships shown in Table 7.

While each of these concepts is defined through logical expression, verbal explanations are missing. They would be needed to be able to map real word situations to these concepts. Although the TOVE framework specifies object libraries and enterprise models as additional components, they remain to be developed.

```
(define-class situation
:axioms
(forall (?a ?s)
        (not (= S_0 (do ?a ?s))))
(forall (?a1 ?a2 ?s1 ?s2)
        (=>     (= (do ?a1 ?s1) (do ?a2 ?s2))
                (= ?a1 ?a2)))
(forall (?s)
        (not (< ?s S_0))
(forall (?a ?s1 ?s2)
        (<=>    (< ?s1 (do ?a ?s2))
                (and    (Poss ?a ?s2)
                        (leq ?s1 ?s2))))
(forall (?P)
        (=>     (and    (?P S_0)
                (forall (?s ?a)
                        (=>     (?P s)
                                (?P (do ?a ?s)))))
                (forall (?s)
                        (?P ?s))))
```

Figure 11. Definition of Situation in TOVE
(www.ie.utoronto.ca/EIL/tove/ontology/basic_sitcalc_theory.html, 5/5/1998)

The TOVE ontologies are an important step forward in the formalization of knowledge about the enterprise. They have been used to formalize knowledge about such core management issues as business process reengineering and activity-based costing. While they are hampered by limitations of first order logic, the difficulties of describing in detail all relationships that hold within a domain, the lack of natural language descriptions and the difficulty of avoiding terminological overlap between the many ontology components, the development of TOVE has yielded many fruitful insights. One widely recognized requirement for the improvement of the ontologies are tools which facilitate the mapping of real world situations to the ontologies.

TOVE, Cyc and Ontolingua server provide examples of an evolution towards more formal descriptions of knowledge and the application of automated reasoning systems. Currently none of the three systems is sufficiently mature to support scientific knowledge and reasoning in the social sciences. However, the systems show the potentials of inferencing systems which have largely been neglected by the other knowledge infrastructures described in this chapter.

Table 7. Partially ordered actions and related concepts defined in TOVE

Class		Function	Relationship		
Activity	Sequence	Initiate	Next_action	Finishes	Before
Poset	Join_action	Terminate	Parallel	After	_delay
_action	Serial	Activity	Leq_sub	During	After_delay
Determinis-	Synchro-	_interval	Successor	Overlap	Equals
tic	nized_join	Split	Initial	_delay	Successive
Total	Synchro-	Join	_subaction	Sometime	_overlap
_ordered	nized_split		Final_	_during	Nonconcur-
Patial_order	Complex		subaction	Start_synch	rence
_action	_poset		Overlaps	ronization	Full_synchr
Sequential				Meets	onization
_poset					Leq_total

Source: www.ie.utoronto.ca/EIL/tove/ontology/poset_def.html, 5/5/1998.

2.3. Visions of scientific knowledge infrastructures

The theoretical and empirical analyses of knowledge infrastructures conducted in the previous sections have some limitations. The theoretical analysis was based on the existing publishing system. It may fail to recognize some potentials for discontinuous change. The limitations of empirical analysis have been described in detail above. To reduce the risk of overlooking novel characteristics which may only emerge in the future, visions of knowledge infrastructures can be examined. While their complete realization may be in doubt, such visions provide novel perspectives on knowledge infrastructures which allow to identify additional potential characteristics.

2.3.1. Hypertext-based visions

In 1945 Bush envisioned an infrastructure which would allow instant access to all knowledge of humankind. He described a device called 'memex' which would allow access to this knowledge and which would keep a trail of the uses of knowledge. He argued that such a device would need to emulate the associative memory of the human (Bush 1945). From a similar perspective, Engelbart (1995) has argued that the focus should be on augmenting the human intellect, collaboration and discursive thinking. He developed the first hypertext system 'AUGMENT' which initially ran on mainframes and subsequently on workstations (Münz 1996).

Influenced by Bush's ideas Nelson coined the term 'hypertext' (Conklin 1987, p 23) for systems which allow "non-sequential writing." (Nelson 1981, p 2) Hypertext became a synonym for a vision of future knowledge infrastructures which are based on fragments of texts that are intensively cross-referenced in combination with a user interface which allows accessing text in a non-linear manner (for an in-depth discussion of hypertext systems see Spring 1991, pp 235-257). The vision is strongly based on the goal of improving the interaction between human and computer and on reflecting the associative structure of human memory in the system. Because of the broadness of the visions, the implementation details of hypertext systems vary considerably. In a review of 18 hypertext systems, Conklin distinguished the following features of hypertext systems (see Table 8):

Table 8. Characteristics of hypertext systems

Characteristic	Description
Hierarchy	Is there specific support for hierarchical structures?
Graph-based	Does the system support non-hierarchical (cross-reference links)?
Link types	Can links have types?
Attributes	Can user-designated attribute/value pairs be associated with nodes or links?
Paths	Can many links be strung together into a single persistent object?
Versions	Can nodes or links have more than a single version?
Procedural attachment	Can arbitrary executable procedures be attached to events (such as mousing) at nodes or links?
String search	Can the hyperdocument be searched for strings (including keywords)?
Text editor	What editor is used to create and modify the content of nodes?
Concurrent multiusers	Can several users edit the hyperdocument at the same time?
Pictures or graphics	Is some form of pictorial or graphical information supported in addition to text?
Graphics browser	Is there a browser which graphically presents the nodes and links in the hyperdocument?

Source: Conklin 1987, p 21.

Hypertext concepts have been incorporated into many software systems. Nonlinear documents are essential concepts in many current groupware systems. The most visible implementation may be the World Wide Web, which however, falls short on several counts of being a full hypertext system. On the World-Wide Web, for example, hyperlinks are unidirectional and usually untyped. There is no support for the collaborative creation of documents. The World-Wide Web, however, provides an additional feature: the inclusion of video and audio documents. Hypertext systems which are capable of supporting such documents are called 'hypermedia' systems.

An additional limitation of hypertext systems is the emphasis on explicit links between texts and on explicit document structures - even if non-linear. Implicit interdependencies - e.g. among articles which have similar attributes may be equally important. As the popularity of search engines on the internet shows, powerful algorithms may identify related articles, even if they are not directly linked.

2.3.2. Seminar

Guedon (1997) proposes several visions of future knowledge infrastructures. He calls his first vision 'the constantly active seminar'. It would allow “dozens of specialists to tune in constantly to the peers and the best laboratories of the planet.” (Guedon 1997, p 81-82) In the seminar, knowledge looses its static properties and arises from discourse among the preeminent researchers in a field. The purpose of the infrastructure is to enhance the flow of communication. Knowledge is created, evaluated, refined, disseminated and applied through communication. The infrastructure stores the trail of discourse and hence preserves a history of the knowledge. From time to time a synthesis of the knowledge may be developed in a parallel encyclopedia-like infrastructure. The focus on communication, however, gives rise to issues such as approaches for the identification of communication partners, mechanisms for generating and enhancing discourse, structures and types of discourse between researchers etc. Another issue is how to handle admission to a seminar. Guedon proposes that mechanisms similar to the selection of editors in traditional journals could be employed. The popularity of discussion groups on the internet shows how much potential this vision has. Members of discussion groups acquire their knowledge from following threads of discussions, from asking members of the discussion group and - to a lesser extent - by reviewing past discussions. Similarly, Harnad emphasizes that “the purpose of scholarly publication is COMMUNICATION -- with peers, and for posterity.” (Harnad 1992) He envisions instant peer-review and argues for the evolutionary model of publishing rather of pursuing a “stationary state that was ready to be etched in stone.” (Harnad 1992)

2.3.3. Eco-museum

Guedon (1997) offers another radically different vision of a knowledge infrastructure. He argues that one of the purposes of a knowledge infrastructure should be to extend the view of the individual beyond his familiar territory. A knowledge infrastructure should include mechanisms to expose an individual to other perspectives and types of knowledge than those which he has come to expect. Much like museums which aim to expose the industrial worker to the life of the farmer and vice versa, this requires a strategy which identifies the perspective taken by individuals and then matches it with other perspectives. This vision focuses on a single characteristic of knowledge infrastructure. It can easily be added to any infrastructure which is able to identify user characteristics.

2.4. Synthesis

The theoretical and empirical analyses of the previous sections have shown several existing and potential kinds of knowledge infrastructures. Moreover, few sharp boundaries seem to exist between different types of infrastructures. While many characteristics have been identified, it is not clear which of them are key. Differences which may seem defining today can vanish quickly as information technology evolves. As was argued above, the focus of the analysis needs to be turned from the level of categories of knowledge infrastructures which eventually may become predominant to the level of characteristics which knowledge infrastructures may exhibit and the relationships between these characteristics. Therefore the subsequent synthesis will focus on the characteristics of knowledge infrastructures and the way in which they can be combined. The identification of stable types is left for the future when the field will have sufficiently matured.

In the previous sections many characteristics of knowledge infrastructures have been identified. They may not be free of overlap. To identify overlap the characteristics can be organized into groups of similar characteristics. This requires the identification of appropriate perspectives and grouping criteria. Subsequently, each characteristic will be assigned to the appropriate group. Finally each group will be analyzed in detail. The resulting set of characteristics is shown in Table 10.

The perspectives which are used for group formation can be based on the main perspectives which have been established for the analysis of information systems. A variety of authors have proposed key perspectives along which to distinguish an information system: In 1987, Zachmann proposed three different perspectives: data, process and network (Zachmann 1987; Sowa & Zachmann 1992). This approach has been refined by subsequent information systems development methodologies. With the advent of object-oriented approaches, the data perspective has been replaced with an object perspective which is concerned with the classes of phenomena, their attributes, methods and relationships. Scheer (1998b) has shown that the network perspective is a special case of a more general organizational perspective which focuses on roles and relationships as well as the objectives of the system.

In the context of knowledge infrastructures, more salient terms can be used for the sets of characteristics which correspond to the perspectives. The characteristics, which are close to the process perspective, for example, describe how participants interact with a knowledge infrastructure. Thus the term 'mode of interaction' is used for this group of characteristics. Characteristics in the object/data perspective are concerned with different aspects of the 'structure of knowledge'. In the organization perspective, characteristics deal with different aspects of 'governance'. Another set of characteristics relates to the ways in which knowledge is presented. While this perspective is not as ubiquitous as the previous three in all phases of information systems development, a presentation perspective is central to many information systems architectures, most notably client-server. Last but not least, a characterization of knowledge infrastructures needs to include technological choices. This leads to five groups of characteristics shown in Table 10: 1) Mode of interaction, 2) Structure of knowledge, 3) Presentation, 4) Governance, 5) Technology. In the following each group of characteristics will be analyzed.

Besides the characteristics which have been identified in the previous chapter, a few other authors have identified key characteristics of electronic journals. In 1997, Eason et al. analyzed the match between electronic journals in different scientific disciplines. Their study had pilot character and made extensive use of secondary data. The researchers distinguished three groups of characteristics: a) amount and type of multi-media, b) structure of the material and search mechanisms, c) support for scholarly communication. All three groups are already included in the subsequent synthesis.

In an analysis of features of electronic journals Gaines distinguished technological and sociological aspects of electronic journals. He identified 14 different dimensions of electronic journals which are shown in Table 9. With few exceptions, the characteristics of both studies are contained in the characteristics described in the currrent and previous chapters. The fourth column in Table 9 maps Gaines' characteristics to the larger set of characteristics shown in Table 10.

Table 9. 14 dimensions of electronic journals identified by Gaines

#	Dimension	Description	Corresponding attribute
1	Private – public	To whom material is available, registration	Governance-audience
2	Discursive – archival	Focus on communication or storage or both	Organization of knowledge
3	Requested – sent automatically	Push or pull or both	Interaction-Notification
4	Moderated – unmoderated	Existence of some form of review process	Interaction-types
5	Standard format – nonstandard format	Standardization of document formats used	Presentation – document format
6	Accessible content – inaccessible cont.	Access only to index, abstract or to full content?	Governance-audience
7	Multimedia – textual	Multimedia support	Organization-types
8	Transient – permanent	Material available for a limited time or indefinitely	Technology – duration of record-keeping
9	Mutable – immutable	Whether authors can change a text (without versioning)	Interaction – updating
10	Authenticated – unauthenticated	Mechanisms for checking whether material is authentic	Technology – document authentication
11	Registered – unregistered	Whether users need to registered	Governance-authentication
12	Indexed – not indexed	Whether material is indexed e.g. in popular search engines	None (this is an external attribute)
13	Annotable – unitary	Whether system supports precise citation	Interaction-types
14	Attributed – annonymous	Whether author is identified	e.g. organization-structuring criteria, governance-authentication

Source: Gaines 1996.

2.4.1. Mode of interaction

Users may interact with a knowledge infrastructure in different ways and different roles. While traditional journals and their electronic counterparts focus on the reader role, other infrastructures provide support for authoring and review.

Table 10. Characteristics of web-based knowledge infrastructures

	Type[1]	Alternatives							
Mode of interaction	Supported types of interaction	Read/browse	Annotate/comment✎	Review	Authoring				
Mode of interaction	✎Online annotation	No	Yes, restricted to specific group (editors...)	Yes, by any reader	Yes, anonymously				
Mode of interaction	✎Annotation shown	No	Yes, shown on original	Yes, link from original to list of comments	Yes, in separate medium (e.g. disc. Forum)				
Mode of interaction	Electronic submission	No	Via email	Via FTP	Via HTTP	Online creation & editing			
Mode of interaction	Time of publishing	Instantly at submission	Immediately at acceptance	After acceptance on publishing date of issue					
Mode of interaction	Updating contributions	No (contributions are immutable)	Direct update (authors update contributions without review)	Reviewed update (review required before or after update)					
Mode of interaction	Email notification	No	Yes, subscription list	Yes, personalized notification					
Structure of knowledge	Structuring criteria	Temporal (e.g. by "volume")	By subject	By author	Other (site, user...)				
Structure of knowledge	Types of content	Abstracts	Papers	Data	Problems	Events	Multi-media	Trails	Other
Structure of knowledge	Link directionality	Unidirectional (web-standard)	Bi-directional						
Structure of knowledge	Types of relationships	Unspecified	Generalization / specialization	Abstraction / summarization	Part / whole	Other			
Structure of knowledge	Maintenance of consistency	Not available	Via manual procedures	Semi-automated consistency checks	Automated consistency checks				
Structure of knowledge	Identification of missing knowledge	No support	Shows list of missing terms (from search inputs)						
Structure of knowledge	System scope	Closed	Open						
Structure of knowledge	Versioning	No versioning	Live publications	Archival publications	Retired publications				
Presentation	Visualization aids	None	Site map (tree)	Graphical node map					
Presentation	Navigation aids	None	"Hit list"	Random node functionality					
Presentation	Filtering mechanisms	None	Extract nodes based on read count	Extract nodes based on ratings, characteristics	"Bubble-up feature"	Other			
Presentation	Internationalization	No	Multi-lingual interface	Multi-lingual interface and content					
Presentation	Customization / personalization	None	User interface	Filtering mechanisms	Stores user profile				
Presentation	Separation of knowledge & presentation	None	Multiple, non-linear presentations of the same knowledge	Dynamic generation of presentations	Context-sensitive presentations				
Presentation	Client configuration	Web browser	Browser with Javascript	Browser with Java applet support	Browser with plug-in	Custom client application			
Presentation	Document format	PDF	PostScript	Html	Plain text	Other			
Presentation	Print version	None	Issues	Volumes					

Table 10. (Continued)

Governance	Selectivity	Inclusive, available to the public			Exclusive; participation tied to qualifications/profile		
	Authentication	Never required		Required for some functionalities⤵		Always required⤵	
	⤷Authentication mechanisms	Username and password	Http cookies		Certificates	Internet address; domain	
	Anonymous contributions	No			Yes		
	Revenue model	Subscription	Pay-per-view	Pay-per-submission	Advertisement	Sponsorship, grant	
Technology	Server type	Html file server	Database	Search engine⤵	Workflow engine	Inference engine	
	⤷Search specification	Limited (e.g. by title, author)	Full text by keyword	Full text by concept	By non-textual criteria	Via concept spaces	
	⤷Relevancy ranking	None	By closeness of match	Based on user preferences	Via network analysis	Other	
	Standards	Java	JavaScript	XML	Active server pages	Dynamic html	Other
	Document authentication	No			Yes		
	Duration of record-keeping	Permanent			Transient		
	Support for sessions	No			Yes		

[1] Triple border to the right of this field: only one of the characteristics can be true at any time; single border: characteristics can co-occur.
⤵ Characterstic has sub-characteristics; ⤷: this is a sub-characteristic
For an earlier version see Hars & Malhotra (1997).

Higher interaction also implies shorter turnaround times for publication. To allow intense discussions, some infrastructures allow instant publication and defer review to after publication. Another approach for improving interaction with an infrastructure is to allow it to take initiative by alerting users to new information that is valuable for them or for automatically requesting them to review new contributions which fits their profile of interest.

2.4.1.1. Type of interaction

Supported types of interaction	Read/browse	Annotate/comment⤵	Review	Authoring
⤷Online annotation	No	Yes, restricted to specific group (editors,...)	Yes, by any reader	Yes, anonymously
⤷Annotation shown	No	Yes, shown on original	Yes, link from original to list of comments	Yes, in separate medium (e.g. disc. forum)

Current knowledge infrastructures differ greatly in the extent to which they allow user interaction. In traditional journals, reading and authoring are separated for technical reasons. Many online journals replicate these limitations of paper technology and target their sites primarily towards the reader. Such sites focus on dissemination and provide no direct support for authors with the exception of displaying submission instructions or accepting files via email.

Information technology permits more advanced levels of interaction such as annotation and on-line discussion of contributions. Only a small number of knowledge infrastructures currently provide this capability. Among the journals which support annotation, some restrict the ability to a subset of the user base, such as reviewers or editors. Others provide this capability for any reader, in some cases even anonymously. An example is Principia Cybernetica Web (pespmc1.vub.ac.be/recent.html). The presentation of annotations differs. Some display the comment or the comment title as part of the document. Others display a link to a list of comments. In some cases, comments can only be accessed from another medium (e.g. a discussion forum) and even their existence is not visible from the source document.

The technology for providing annotation features and discussion lists has matured rapidly. Such capabilities can now be added easily and reliably to any web-based infrastructure. It is not clear, however, to what extent knowledge infrastructures will adopt these features. Some journals have found that their annotation features are rarely used. MISQ Discovery (www.misq.org/discovery), for example, has discontinued a discussion board it had initially set up for their articles because of apparent lack of interest. The problem may be the low number of annotations which a typical scientific article may receive. This presents additional challenges for the functionality which are very different from traditional discussions lists. It may be crucial for example, to alert the author via email when an annotation has been added. In addition, the ability to aggregate annotations and to identify those articles which have lead to discussions may be important. Structured comments which allow an evaluation of an article according to predefined criteria may be more useful than the ability to provide unstructured comments.

An increasing number of electronic journals provide support for the review process (Barua, Chellappa & Whinston 1996). This may include a separate space for papers under review, to which either all readers or just the reviewers have access. Some infrastructures incorporate software for managing the review workflow.

The fourth interaction type is authoring. It comprises the creation of new knowledge as well as the synthesis and integration of existing knowledge. Support for authoring requires the existence of online editing mechanisms. Although some journals such Ontolingua server and the Information Systems Cybrarium provide editors, the vast majority of electronic journals do not support this function; they require the submission of a finished article.

2.4.1.2. Electronic submission

Electronic submission	No	Via email	Via FTP	Via HTTP	Online creation & editing

An increasing number of electronic journals accept electronic submission of articles. An exception are hybrid journals which often continue to require paper copies. Email currently is the prevalent approach for online submission. In most cases such submissions are processed manually although some infrastructures automatically publish the content of email submissions. Some journals provide an FTP server to which authors can upload their articles. Usually this needs to be accompanied by an email notification. Submission via FTP is problematic when a contributions consist of multiple files. Few journals allow submission via the world wide web's HTTP protocol and even fewer allow the online creation / editing of submissions.

2.4.1.3. Time of publishing

Time of publishing	Instantly at submission	Immediately at acceptance	After acceptance on publishing date of issue

Most journals publish contributions after successful review at the time that the issue is published. An increasing number of journals publish articles as soon as they have passed the review process. Thus readers may be able to 'preview' an accepted article before the official release date of the issue. This practice indicates a breakdown of the traditional paradigm. Traditional electronic journals have introduced this feature because it became obvious that there is no need to wait for publishing an accepted article until all articles have been reviewed. This does not address the fundamental issue, however: the obsolescence of structuring an online journal into issues.

Journals which provide electronic support for the review process may publish a submission instantly in a separate review section where access is restricted to reviewers and editors. Upon acceptance, the article is moved to the public area. Other journals have started to reengineer the review process and to parallelize publishing and review. At ISWorldNet (www.isworld.org/isworld.html), for example, new submissions for a topic are published immediately, although in a separate section (Ives & Zmud 1994). Reviewers as well as the audience are invited for comments. If the review panel comes to a negative decision, the submission is removed from the site (see also proposition 10 in Table 5 on page 20 and the related discussion).

2.4.1.4. Updating contributions

Updating contributions	No (contributions are immutable)	Direct update (authors update contributions without review)	Reviewed update (review required before or after update)

In the paper environment, authors are not able to change their contributions after publishing. Although this limitation does not persist in the online environment many online journals do not allow changes. Among the few infrastructures which allow changes, changes may or may not need to be reviewed. This may occur either before or – as in the case of Stanford's Internet Encyclopedia of Philosophy - after the change has been made. Unreviewed updates may be advantageous in highly collaborative environments which require the timely exchange of ideas.

2.4.1.5. Notification

Email notification	No	Yes, subscription list	Yes, personalized notification

Another aspect of interaction is the distinction between active and passive electronic journals. Most electronic journals expect the reader to take initiative for retrieving information. Some electronic journals, in contrast, push information to the reader via email notification. The publisher Springer (www.springer.de), for example, offers a free 'link alert' service which emails tables of contents of new journal issues to interested researchers. The tables of contents include links to the full text which, however, is accessible for subscribers of the journal only. The Stanford Encyclopedia of Philosophy automatically notifies the corresponding editor if a document is changed or a new document is added. In the Information Systems Cybrarium an email is automatically generated to notify the authors when a reader adds a comment to one of its documents. Because researchers utilize knowledge infrastructures asynchronously notification mechanisms are valuable for coordinating activities. Different degrees of notification functionality need to be distinguished. The sophistication of notification functionality may range from simple subscription lists where the same alert is sent to each member of the list to personalized notifications. In the latter case, the reader may select combinations of topics, keywords, authors, assign weights and determine the frequency of notification. Alternative variants of notification may coexist in a single knowledge infrastructure. In some parts, predefined notification may be available, while another area may provide customizable email alerts. An alternative to notification via email is to notify a user of important events when she logs on. This can only be effective once an infrastructure has a high level of repeat users.

2.4.2. Structure of knowledge

This second dimension refers to the problem of organizing knowledge and ensuring that over time a consistent, integrated body of knowledge emerges. The degree to which knowledge is organized is determined in large part by the meta structure on which the knowledge infrastructure is based. It may be available explicitly or implicitly. While online infrastructures which incorporate a database require an explicit model for technical reasons, infrastructures which are based on static html pages often lack an explicit conceptual model. Even in the latter case, a clear definition of the meta structure becomes important as scope and functionality grow. Several methods for defining such meta structures are now available (Isakowitz, Stohr, & Balasubramian 1995).

2.4.2.1. Structuring criteria

Structuring criteria	Temporal (e.g. by "volume")	By subject	By author	Other (site, user...)

An important aspect of the organization of a knowledge infrastructure are criteria for structuring knowledge. Many electronic journals still adopt the traditional paper-centric structuring into volumes and issues, although contributions within an issue may have nothing in common except the publication date. More advanced electronic journals replace or complement this approach with a subject tree or a network of categories which the reader can browse. This enables the reader to easily identify contributions belonging to similar topics. Category networks are not limited to subject trees; for example, author/affiliation networks, lists or graphs of synonyms, etc. can be added.

2.4.2.2. Types of content

Types of content	Abstracts	Papers	Data	Problems	Events	Multi-media	Trails	Other

Infrastructures differ in the type of content which is accessible to the reader. Online journals which are modeled on traditional journals typically separate pages with abstracts from pages which contain full text. While the WWW inherently supports multimedia data such as moving images and sound, most current infrastructures do not exploit this feature. Hybrid journals don't allow multimedia content. Some journals introduce pages for special elements of scientific knowledge such as data, events and problem descriptions. Another aspect is support for 'trails'. A trail (or path) is a persistent set of links which allows a reader to navigate in a predefined way through a specific set of nodes. Trails can be used to create compilations of knowledge. Instead of having to decide which way to go at every link, this allows a reader to 'stay on a trail' and systematically explore a collection of documents (Shipman et al. 2000).

2.4.2.3. Link directionality

Link directionality	Unidirectional (web-standard)	Bidirectional

Another source of differences lies in the treatment of relationships. On the world-wide web links (e.g. cross-references) are unidirectional. From the source of a hyperlink, it is possible to navigate to the destination URL. It is not possible, however, to move from the destination of a hyperlink to its source because the hyperlink is only embedded in the source document. This is similar to the paper environment, where a citation identifies another document; the referenced document, however, provides no information about the referencing document. In contrast, bi-directional relationships allow traversing a link in both directions. When a document is referenced it becomes possible to navigate backwards from the referenced document to the referencing document. This has important implications: It is possible to follow the evolution of an idea - from the original idea to documents which cite the idea and then to documents which cite these documents etc. It becomes immediately evident how many times a document has been cited. In addition, the detection of similarities is simplified: an author can check which other articles extend a certain idea. This may reduce the duplication of research efforts and may increase collaboration between researchers.

2.4.2.4. Relationship types

Types of relationships	Unspecified	Generalization / specialization	Abstraction / summarization	Part / whole	Other

Similar to the case of nodes, different types of relationships need to be distinguished. On the world-wide web, most relationships have no type - although it is technically possible to assign names to relationship types. While online knowledge infrastructures distinguish many different relationship types, some types are likely to occur frequently: a contains relationship - e.g. when a concept is created by joining a number of other concepts. An example is the breakdown of the concept 'computer' into the concepts 'hardware' and 'software'. A second concept is the division of a concept into subconcepts - e.g. the structuring of 'software' into 'operating system', 'database', 'networks' and 'application software'. A third relationship is the abstraction/summarization relationship. It occurs, for example, between an article and its abstract. Many other relationships are possible.

2.4.2.5. Maintenance of consistency

Maintenance of consistency	Not available	Via manual procedures	Semi-automated consistency checks	Automated consistency checks

Besides these static aspects of the organization of knowledge, issues related to the change of knowledge need to be considered. When new knowledge is inserted or existing knowledge is updated, problems of consistency arise: does the change invalidate, duplicate or contradict existing knowledge? In a paper-based environment authors and reviewers attempt to ensure a minimum level of consistency, but the process is manual. In an electronic environment, semi-automatic or automatic mechanisms may exist to detect inconsistencies.

2.4.2.6. Identification of missing knowledge

Identification of missing knowledge	No support	Shows list of missing terms (from search inputs)

This characteristic can be found at the Free-Online Dictionary of Computing (wombat.doc.ic.ac.uk). It has a feature that promotes contribution of knowledge from visitors. When visitors search for a term that is not available in the dictionary, the missing term is automatically added to a list of missing terms. The system records the frequency of such requests and displays the most frequently reported missing terms on a separate page. In this way contributors can be directed towards areas where there is sparse knowledge. In current IS-related knowledge infrastructures support for such presentation features is very limited.

2.4.2.7. Scope

System scope	Closed	Open

Another issue is the scope of the knowledge infrastructure. Most current electronic knowledge infrastructures operate on the assumption that all articles will be physically stored on their site. However, location is not important on the internet (see proposition 4 in Table 5 on page 20). Some infrastructures include articles independently of where they are stored on the world wide web. Technically this can be achieved by linking to an external page through an HTML-frame (an example is shown in Figure 29 on page 192). The first type of knowledge infrastructure is closed; it has the advantage of full control over the articles. Open knowledge infrastructures, in contrast, have the disadvantage of uncertainty about the continued availability of the article. However, they can create a larger and more comprehensive knowledge base as they no longer require ownership of all articles.

2.4.2.8. Versioning

Versioning	No versioning	Live publications	Archival publications	Retired publications

Articles in paper journals are static. In the electronic environment, articles can be changed easily and it becomes necessary to develop an approach for versioning. The simplest approach which, for example, has been implemented at the Stanford Encyclopedia of Philosophy, is to replace an older version with a new version. MISQ Discovery distinguishes between archival publications which are not updated (except for spelling errors etc) and live publications which include a versioning mechanism. Retired publications remain available but are no longer valid or relevant. A major problem related to versioning is handling relationships and notification needs.

2.4.3. Presentation

Scientific knowledge grows rapidly. Over the last 300 years, the number of articles has doubled approximately every 15 years Price (1986, p 6). For an analysis of specific growth rates in various disciplines see Menard (1971). Keeping up with the literature is a challenge for scientists in every field. One approach for reducing the cognitive load on the scientist is to improve the presentation of knowledge. Presentation mechanisms may focus both on adequately presenting the internal components of an article as well as the web of relationships with other pieces of knowledge into which an article is embedded.

2.4.3.1.Visualization aids

Visualization aids	None	Site map (tree)	Graphical node map

Many sites provide sitemaps - trees which graphically display the node structure. The approach breaks down in extensively cross-referenced sites which have a network rather than a tree structure. Several approaches for more refined node maps are described below in the section on search technology.

2.4.3.2. Navigation aids

Navigation aids	None	"Hit list"	Random node functionality

Besides visualization aids, browsing can be simplified with navigation aids. Some sites display a 'hit list' containing, e.g. the best-rated or most frequently accessed articles. An example is the Public Access Computer Systems Review (info.lib.uh.edu/pacsrev.html) which maintains a page that allows the reader to jump directly to the ten most frequently accessed articles. Similarly Principia Cybernetica Collaboratory has a 'HIT PARADE' page which displays knowledge nodes according to their popularity. Another navigation aid which supports casual browsing is a random node functionality. Both the Free Online Dictionary of Computing and Principia Cybernetica allow the user to access a node randomly. This feature provides a starting point for the users to explore the knowledge base even if they are not particularly aware of the knowledge base contents.

2.4.3.3. Filtering mechanisms

Filtering mechanisms	None	Extract nodes based on read count	Extract nodes based on ratings, characteristics	"Bubble-up feature"	Other

The hit lists discussed in the previous section are a simple version of filtering mechanisms which extract a subset of nodes from a knowledge base. Filtering mechanisms rely on node characteristics such as access counts, ratings, types etc. in conjunction with reader preferences to return a subset of nodes which is highly relevant for a specific reader. Filtering mechanisms may be predefined or customizable by the user. A more sophisticated approach is used in some help desks. 'Bubble up' technology (El Sawy & Bowles 1997) ensures that highly relevant articles gradually are moved to the top of a filtering mechanism.

2.4.3.4. Internationalization

Internationalization	No	Multi-lingual interface	Multi-lingual interface and content

Another aspect of presentation is support for different languages. With few exceptions such as the Electronic Journal on Networks and Distributed Processing (rerir.univ-pau.fr/index.html) and the Information Systems Cybrarium, most current knowledge infrastructures only support a single language. To achieve full internationalization, an infrastructure needs to provide interfaces in multiple languages. Content must be accepted in multiple languages and it must be possible to restrict the presentation, search and navigation of documents to the subset of the languages which a reader is familiar with.

2.4.3.5. Customization

Customization / personalization	None	User interface	Filtering mechanisms	Record of interactions

Advanced information systems allow users to customize the interface. Users may select how information is presented to them by default or what default mechanisms are to be applied for filtering information. This information requires knowledge infrastructures to maintain user profiles. Very few knowledge infrastructures allow such customization. Examples are Ontolingua Server and the Information Systems Cybrarium. Customization can be further improved by recording certain user interactions with the goal of implicitly determining user preferences.

2.4.3.6. Separation of knowledge and presentation

Separation of knowledge & presentation	None	Multiple, non-linear presentations of the same knowledge	Dynamic generation of presentations	Context-sensitive presentations

Paper-based environments significantly limit the presentation of knowledge. Paper is static. The presentation of knowledge is fixed by the author and knowledge is presented to every reader in exactly the same way. In contrast, information technology is able to separate the presentation of knowledge from its underlying formulation. Separation between presentation and implementation is widely practiced in current information systems and is fundamental to client-server architectures. For knowledge infrastructures, the following aspects of the separation of knowledge and its presentation need to be distinguished:

Multiple, non-linear presentations of the same knowledge: The online environment eliminates the familiar restriction that knowledge must be ordered sequentially. An article may consist of many elements which are linked together in a network of hyperlinks. It may provide several entry and exit points for the reader. It may contain different lines of arguments which are orientated towards the level of expertise and the interest of the reader. This provides additional challenges for knowledge infrastructures. Authors must write their articles in a way that different modes of reading and different reading paths are possible. Knowledge infrastructures need to provide mechanisms for guiding a reader along his specific path and to hide unnecessary detail. For example, an author may mark certain parts of his article as detailed discussions which are targeted mainly towards fellow experts. A knowledge infrastructure could then hide these parts automatically from the casual reader. Additional mechanisms are needed which allow the reader to monitor his position with respect to the overall body of knowledge. They need to ensure that all relevant elements have been traversed when moving through a knowledge network. In the current implementation of the world-wide web, there is little functionality to support this. The reader is kept informed about links already visited. However, he is not able to track which nodes he still needs to visit when reading a specific domain. The move towards non-linear presentation of knowledge also requires rethinking traditional three-tiered article structures (e.g. introduction, main body of knowledge, conclusion).

Dynamic generation of knowledge presentations: The same knowledge can often be presented in different ways. Various types of diagrams can be used to present conceptual models. Tables or graphs may show dependencies between data. With sufficient information systems support, alternative representations can be generated automatically from underlying text and data. This allows to assemble the data in the format which is preferred by the reader. Finally, it enables the reader to view additional representations which the author may not have thought about. Data, for example, may be used for analyses with a purpose that is different from the original study.

The ability to change the way in which the content is presented dynamically to the needs of the reader changes the responsibilities of the author. The author may be expected to invest additional effort to increase the flexibility in which the content can be viewed. This may include describing the data in further detail so that other researchers may view it for other purposes. The data collected, therefore, may be treated as knowledge in its own right instead of serving only as evidence for one specific project.

Context-sensitive presentation of knowledge: Information technology allows to tune the presentation of knowledge to the requirements of the reader and context. A student, for example, needs to approach the same body of knowledge in a different way than a practitioner. A reviewer, will need to focus on other details than a researcher who is examining the progress of his colleagues. For the research community - and IS research - this requires the formulation of detailed, possibly dynamic, user models.

2.4.3.7. Client configuration

Client configuration	Web browser	Browser with Javascript	Browser w. Java applet support	Browser with plug-in	Custom client application

Presentation functionality is influenced by the software configuration of the client. A basic web-browser can only present pages on the screen and requires all presentation-related interaction to be processed on the server. Browsers with support for a scripting language (e.g. Java Script) allow more flexibility in the presentation. The three other configuration types further increase the sophistication of presentation; their disadvantage are delays for downloading applets and installing plug-ins or custom client applications.

2.4.3.8. Document format

Document format	PDF	PostScript	Html	Plain text	Other

On the internet support for document formats varies widely. Among electronic journals, Adobe's PostScript and Portable Document Format (PDF) are very popular. PostScript has the disadvantage that it cannot be viewed on screen without additional software. The PDF format ensures that documents can be printed easily and faithfully. However, PDF requires a separate plug-in and is not well integrated with web technologies (e.g. Java, JavaScript). An additional disadvantage of PDF is that the creation of PDF files requires additional software while most current word processors can generate PostScript and HTML files. As display technology matures and the share of articles read on slim, lightweight LCD displays or electronic ink increases (see the discussion on maturing display technology on page 24 above), HTML or XML documents are likely to become the document type of choice.

2.4.3.9. Print version

Print version	None	Issues	Volumes

This attribute determines whether the knowledge infrastructure has a print version. Many online knowledge infrastructures make their content available in print form. In hybrid journals, print and online versions are nearly identical. Both are typically organized by issue. To maintain backward compatibility with the current paper based publishing system several original online journals have chosen to publish a print version of their articles. An example is the Journal of Artificial Intelligence Research which publishes an annual print volume. As online infrastructures mature further, it is likely that print and online versions will diverge more and more.

2.4.4. Governance

This dimension of knowledge infrastructures refers to the organizational and procedural environment in which it is embedded. A crucial issue is the nature of the community. Are the users an anonymous group or do they act as clearly identifiable individuals? A second aspect is the approach taken to recover the costs of operating the infrastructure.

2.4.4.1. Selectivity

Selectivity	Inclusive, available to the public	Exclusive; participation tied to qualifications/profile

This attribute differentiates between knowledge infrastructures which limit access to a specific community and those which provide access to the public. Currently scientific journals are available to anyone interested and willing to pay. Even professional societies such as IEEE or ACM which restrict their membership offer subscription to non-members, although at a higher rate. In an online environment which allows a more active discourse and where the roles of authors and readers converge, however, it may be more important to restrict access to qualified persons.

2.4.4.2. Authentication

Authentication	Never required	Required for some functionalities↬	Always required↬

Infrastructures differ in their requirements for authentication. If they don't charge for their content, they usually don't require authentication to retrieve information. This simplifies access to the system which is particularly important for infrastructures which are accessed infrequently. However, it eliminates the ability to create user profiles which are needed for discursive functionality and collaborative filtering. Many knowledge infrastructures utilize a layered approach where basic browse, search and - in the absence of fees – retrieval functionality do not require authentication. Authentication is only required for more advanced functionalities such as annotation, authoring and review.

2.4.4.3. Authentication mechanisms

Authentication mechanisms	Username and password	Http cookies	Certificates	Internet address; domain

Knowledge infrastructures use different approaches for authenticating their users. The most common approaches are authentication via username and password or via HTTP cookies. Authentication via username and password has the disadvantage of raising an additional barrier for participating in a knowledge infrastructure. HTTP cookies – when used without passwords – have little security; in addition users which share the same browser cannot be distinguished from one another. Certificates provide a more robust form of authentication. The disadvantage is that all participants need to agree on a certificate authority. Some online journals use internet addresses or domains for authenticating groups of users. This is useful for providing access to specific groups (e.g. students from a particular university). It is problematic for identifying individual users as IP addresses may change from session to session.

2.4.4.4. Anonymous contributions

Anonymous contributions	No	Yes

In scientific discourse, anonymity is sometimes desired. In the review process, anonymity reduces the burden on the reviewer, allowing her to make more direct and critical comments. Some infrastructures allow contributors to hide their identity if desired. An example for an infrastructure which allows anonymous comments on articles is Principia Cybernetica Web (see section 2.2.4.2. on page 46 above.

2.4.4.5. Revenue model

Revenue model	Subscription	Pay-per-view	Pay-per-submission	Advertisement	Sponsorship, grant

A key problem for all knowledge infrastructures is the approach taken for covering the infrastructure's costs. Four fundamental approaches need to be distinguished for the generation of revenue:

Subscription: This is the traditional model where the reader or his institution pays an annual subscription fee. The disadvantage is that this approach is economically sub-optimal. Readers who would be willing to pay more than the - very low - marginal costs of viewing but less than the subscription fee cannot access the article. Publishers of electronic libraries such as ACM, Elsevier and Springer have implemented subscription schemes.

Pay-per-view: The reader pays a fee every time that he accesses a document. This can be done via micro-payment or by deducting from an account that needs to be refilled when depleted. Some practical problems remain - such as whether the user will have to pay again when he accesses the same document a few hours later. Another problem is how to increase the chance of retrieving content the reader is looking for without giving away the information in the search process. But the largest problem is determining an adequate price. As articles are retrieved with very different motivations – some are only browsed while others are recognized as being key to a research effort, it is doubtful whether a uniform price would be viable. If the uniform price were oriented towards browsing, it might not generate sufficient revenue. If it were oriented towards papers that are key to a research effort, the number of retrieved articles might be too small. A more promising strategy would be to establish different prices for different articles. Fishwick, Edwards & Blagden (1998), for example, propose to charge only for articles above a certain level of quality while other articles could be retrieved for free.

Pay-per-submission: Another alternative are submission fees. In the Seventies, many scientific journals required authors to pay such fees (Odlyzko 1997). Although this approach would provide an incentive for authors to submit articles of higher quality, acceptance by the authors is unlikely.

Advertising: Advertisements can be added to online infrastructures easily. Reader profiles can be used to target ads individually as well as for additional marketing efforts. While knowledge infrastructures with a very large reader base might prove attractive, it is doubtful whether small communities could be supported through advertising alone.

Free access: The academic community has always sought to create knowledge as a public good. Many authors call for free access to knowledge on the internet. They point out that even today subscription pays only for a small part of the total cost of an article. Authors, reviewers and many editors are not directly compensated for their work. They are paid through research grants, their research institution etc. Odlyzko, for example, argues that "Scholarly publishing is a public good, paid for largely (although often indirectly) by taxpayers, student's parents, and donors." (Odlyzko 1997) Knowledge management infrastructures could be funded in similar ways. The advantage would be that no barriers would exist for access to knowledge.

Knowledge infrastructures typically use a combination of these four strategies. For example, access to the ACM Digital Library which contains the full text of 21 journals since 1991 is available for an annual subscription fee of $180 for ACM members. Alternatively, any reader can purchase individual articles for a fee of $10 per article. It is possible to pay online via credit card. Browsing and searching are free (www.acm.org/dl/slide8.html, 6/18/1999).

The identification of a suitable approach for generating revenue is a key problem for the success of knowledge infrastructures. The shift towards electronic knowledge infrastructures has raised considerable concern about the impact on revenues for commercial publishers and scientific societies, which often make significant profit from journal publishing. In traditional journals, revenue is generated primarily through subscription fees. In scientific journals, advertisements and the sale of individual copies generate only a small proportion of total revenue. Journals use price differentiation; an institutional subscription costs typically 3 to 4 times as much than an individual subscription (Fiswhick, Edwards & Blagden 1998). In the period from 1985 to 1996, the Blackwell Periodicals Price Index for scientific journals has risen twice as fast as the consumer price index (Fiswhick, Edwards & Blagden 1998). This 'serials crisis'(Arnold 1993, p 74; Cummings et al. 1992, pp. 83-99; Okkerson 1996, pp. 188-192) forced libraries to significantly reduce journal subscriptions. Tenopir and King estimate that the average cost per reading (adjusted by inflation) has increased from \$50-\$60 in 1977 to about \$60-70 in 2000 (Tenopir & King 2000). Could electronic knowledge infrastructure reverse these trends of increasing costs?

Several authors have compared the costs of traditional and electronic journals. Fishwick et al. estimated costs for different components of the publishing processes in traditional and electronic environments (see Table 11). They identified copying and distribution costs as primary sources for savings and argued that the costs of editorial work, composition and marketing are less likely to change. Because of the need for multiple formats, the costs of composition may even increase. Fisher (1997), reporting on the launch of electronic journals by MIT Press, confirmed the reduction in direct costs. Tenopir and King (2000), however, argue that

> '...most publishing activities and corresponding costs are common to both traditional paper-based and electronic media. The small savings from eliminating printing and mailing are offset by increased investment in and costs associated with technology.' (Tenopir & King 2000)

Odlyzko (1997) estimates the average compensation per article which a traditional publisher receives from the scientific community at approximately \$4000. He compares it to the costs of Ginsparg's non-profit physics pre-print archive which processes about 20.000 papers per year. With composition done by the authors and computer costs (including server administration) of not more than \$100.000 annually, he calculates an average cost per article of less than \$5. This may be a lower bound as it does not include the costs for mirror systems in different parts of the world.

Disagreement about potential cost reductions may be rooted in different visions of knowledge infrastructures. Fishwick et al.'s estimate, for example, assumes that the publisher is responsible for layout and copyediting while Ginsparg leaves this responsibility with the author. An electronic journal which is closely modeled on traditional journals, therefore, will have a different cost structure than an archive or a collaboratory.

In an electronic environment, storage and distribution costs are nearly zero (Braman 1995, p 46). Text of a typical journal article is smaller than 100kBytes; if figures are included the average size should not rise above 500kBytes. Using an internal hard drive this leads to storage costs of less than a tenth of a cent per article in 2003 with significant potential of further decrease. Hardware is a major cost factor. A low-end dual processor server is capable of serving 10 articles per minute, which corresponds to a sustained load of approximately 100 kBytes per second. For hardware costs of approximately $3000 this allows the distribution of 14000 articles per day under ideal load conditions. This is enough to support a community of 10,000 readers which each access one article per day. Assuming access to a T1 internet connection at about $700 per month, such a system would lead to annual costs of less than $1 per reader excluding server administration costs. While the costs increase proportionally in smaller communities they may not need to operate a dedicated server. Small communities can often find a member who is willing to dedicate a part of an existing server provided the load is not too high.

Table 11. Illustrative annual cost of quarterly journal (200 pp. 23.5 x 15.5 cm) in UK £

	All print on paper	All electronic	50% split
Origination or first copy costs (fixed)			
- Editorial work	55,000	55,000	55,000
- Composition	35,000	40,000	40,000
- Marketing, promotion, sales overheads	15,000	20,000	20,000
Total	105,000	115,000	115,000
Incremental costs (avg. per copy)			
- Physical distribution	8.00	1.00	4.50
- Physical reprint	8.00		4.00
Total cost at distribution numbers			
1000	121,000	116,000	123,500
2000	137,000	117,000	132,000
5000	185,000	120,000	157,500
10000	265,000	125,000	200,000

Source: Fishwick, Edwards & Blagden 1998.

In the current phase, the predominant revenue models for scientific knowledge infrastructures are subscription and sponsorship (free). An increasing number of journals offer pay-per-view. However, no significant share of revenue is currently generated in this way.

2.4.5. Technology

2.4.5.1. Server types

Server types	HTML file server	Database	Search engine	Workflow engine	Inference engine

Online knowledge infrastructures combine various kinds of server functionality. All infrastructures include a web server which accepts HTML requests from the internet. In its simplest form, the web server translates requests into addresses on the server file system and then returns the files. While file servers can be set up quickly, they are difficult to maintain when the size of an infrastructure increases. In addition, their functionality is limited. An alternative is to link the web server to a database. In such cases the web server translates a request into a query which is run against the database. The result of the query may be further processed to generate an appropriate layout before it is passed on to the client. This increases the flexibility for extracting and presenting content. In addition it provides better support for access control, data administration and backup. Many online infrastructures now rely on databases. Another component which is frequently found is a search engine. Characteristics of search engines will be further discussed in a separate section below. In rare cases, a workflow engine has been added to a knowledge infrastructure. Workflow is useful, for example, for the review process, where tasks need to be coordinated between editors, reviewers and authors.

2.4.5.2. Search technology

Search specification	Limited (e.g. title, author)	Full text by keyword	Full text by concept	By Non-textual criteria	Via search spaces
Relevancy ranking	None	Based on user behavior	Based on network analysis	Other	

The rapid evolution of the internet has lead to the development of sophisticated search engines. The size of the "indexable web" has been estimated at 2 billion pages (Sirapayan et al. 2000). This is a rapid increase over 320 million web pages estimated in 1998 by Lawrence & Giles (1998). Technology developed for search engines such as Google, Altavista, Northernlight which cover a large part of the internet is rapidly being translated into search technology which can be used on individual web sites. In mid-2001 the largest search engine (Google) maintained an index of 1.3 billion web pages (Sullivan 2001). In the following search technology will be analyzed from the perspective of search engines targeted towards the whole internet. The same characteristics are relevant for site-specific search technology, although on a smaller scale.

Two different types of search services need to be distinguished: directory services and search engines. While most search services originated with one of the two types, most current search services now combine aspects of both types.

2.4.5.2.1. Directory service

Sites in this category develop an abstract representation of the WWW in the form of a 'classified list.' (Schwartz 1998, p 974) In its simplest form, it may be a hierarchical structure of subject areas into which web sites are mapped. Readers can use the structure in similar ways as the yellow pages to determine which sites may be relevant to their particular problem. The most successful exemplar of this category is Yahoo (www.yahoo.com). It was started in 1994 by two Stanford Ph.D. students who looked for better ways to keep track of their personal interests on the internet. They chose a fine-grained hierarchical structure which proved to be very useful to the internet community. Although Yahoo has diversified significantly, this structure remains at the core of Yahoo's business. More primitive examples of directory services, are link lists which direct readers to other pages of interest on the internet. Directory services need not be hierarchical. Concept networks could be provided to navigate the web. In contrast to search engines, directory services provide an organized space of concepts which the user can browse to find a search result.

2.4.5.2.2. Search engine

Search engines continuously collect information about documents on the internet. A typical search engine consists of two components. A robot (also called spider or web-crawler) is a software agent which navigates the web and retrieves its documents. Properties of each document such as URL, title, key words, full text, etc. are collected and subsequently an index is built. By following most links on a document, robots can rapidly access large parts of the internet. Altavista's robot, for example, is capable of retrieving 10 million documents per day.

Although several search engines maintain very large indexes, no search engine indexes all pages on the internet, not even all static pages. At issue are not just the resources required for collecting (and updating) the index. A larger number of documents aggravates shortcomings of current search algorithms; more documents introduce more noise into search results. Therefore most search engines only index some core sites fully and sample the rest of the internet (Sullivan 1998).

In the more controlled environment of a scientific knowledge infrastructure, robots may not be necessary. When all information of a knowledge infrastructure is stored in one database, then the database system may automatically perform indexing as content is added, changed or deleted. Robots are needed in most other cases, however. This is especially the case, when information is dispersed among many sites or when contributions are stored as files on a web server.

The second component of a search engine is the query processing system. A user enters the desired properties of a web page - usually a combination of key words and characteristics such as language or publishing date. The query processor then retrieves the matching URLs. Some online knowledge infrastructure provide only minimal search functionality which is limited to title words and names of authors. Full-text search, however, is increasingly common. Advanced query systems allow to specify inclusion or exclusion of specific words. Other features are adjacency searches, e.g. where a word needs to occur in the proximity of another word or concept searches. Depending upon sophistication, the latter use stemming rules to avoid variations in endings. A search for the word 'orders', for example, might return web pages that contain 'order', 'orders', 'ordering' and 'ordered'. More complex concept searches retrieve synonyms. A query for 'client', for example, would also return web pages with the word 'customer'. Query results are returned as lists. Advanced search engines provide a combination of the text query features shown in Table 12.

Despite these technologies, web searches continue to be frustrating and time-consuming. Brin & Page (1998) report that out of the 10 top search engines, only 1 lists itself among the first 10 entries in the result list. Some search engines aim to improve the search result by computing a relevancy ranking. This is based on an estimation of how closely the result matches the query formulation. A result is deemed closer, for example, if the query words occur in the title rather than in the body text, if all words from the query string are found, if the search words are next to each other rather than distributed through the text etc.

Recently, several new technologies have been developed to increase the relevancy of search results. HotBot now incorporates DirectHit (www.teoma.com) a technology that uses selections made by previous users to identify highly relevant sites. The technology records which sites are selected from a result list and how long a user remains at the selection before returning to the result list. If the user returns quickly, this is an indication that the site was not valuable to the user. DirectHit then reduces the relevancy rating for the site.

Table 12. Query features of search engines

Category	Description
Boolean search	Concatenate terms with AND, OR, NOT; In some cases with nested parentheses
Required terms	Specification of terms which must or must not be present;
Truncation	(Terminal and internal) or inhibition of automatic stemming
Exact phrase match	Restricts search result to exact phrase
Proximity searching	Search term may be separated by a certain number of words
Fielded search	Based on markup tags identifying title text, meta text, heading text, link, etc.
Case sensitivity	Search terms are case-sensitive
Category restrictions	Restrictions based on date, domain, language, file type etc.
Concept search	Includes synonyms (e.g. from a thesaurus)

Source: Adapted from: Schwartz 1998, pp 975-976.

The founders of Google have developed another approach for determining the relevancy of a site (Brin & Page 1998). They calculate a page rank for each page on the web by analyzing the network in which a page is embedded. In analogy to citation counts used for assessing the value of scientific articles, page rank depends on the number of links which point to the page. The links are weighted by the page rank of the source pages. Thus a page can be ranked highly because it is referenced by many sites or because it is referenced by few pages (e.g. Yahoo) which have very high page ranks themselves.

One of the problems of searching is the large number of documents which may be relevant for a query. Users are reluctant to scan more than the top 10 entries in a result list (Brin & Page 1998). Therefore much research has focused on providing visual aids for exploring larger result sets. These visual aids provide an organized space of concepts which can be used for further refining a search. An example for such a concept space provided by the Altavista search engine is shown in Figure 12 for the term 'knowledge management'. Altavista identifies concepts which are often associated with 'knowledge' and 'management' and correlations between these concepts. The user can then select specific terms to include or exclude from the search using drop-down list boxes. An example is shown in Figure 12 for the term 'organizational'.

Visualization can be utilized to improve search specification as well as interpretation of search results. Figure 12 shows an example of the former. Several search technologies have been developed to improve the representation of search results. Apple, for example, has developed an approach for creating three-dimensional representations of concept spaces (Guha 1998). Although large graphs can be displayed as networks, they are difficult to use for humans. Several techniques have been developed which distort the network to allow focus in one area while providing visual cues about the rest of the network. As the user navigates the network focus follows. Examples of these techniques are 'fisheye views' and 'moving wall' diagrams (Lei & Danaher 1997).

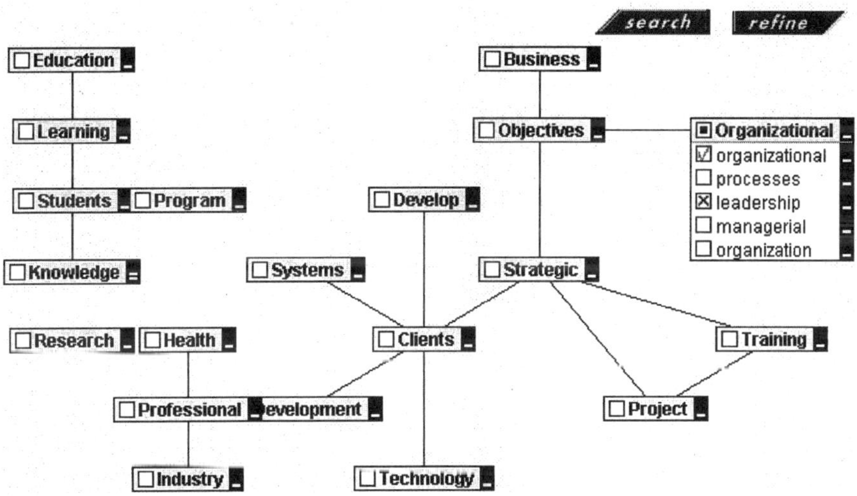

Figure 12. Graphical refinement of search for 'knowledge management' at Altavista Source: http://www.altavista.digital.com, 8/28/1998.

2.4.5.2.3. Novel approaches

A limiting factor for search engines is the inability of the user to give a clear description of what he is looking for. In most cases, a user searches for documents, because he is not knowledgeable in the topic. Searching, thus is a process of sense-making and learning. A user, who becomes interested in 'knowledge management', for example, may start out with the vague idea that it may be relevant for the design of a new information system. He then finds that much of knowledge management deals with tacit knowledge which resides in employee's minds and in social networks and decides to reengineer his business processes accordingly. Insights taken from a search, therefore may significantly differ from what the user initially expected. There are cases, of course, where the user has precise search objectives. Examples are searches for flight schedules or information about a commercial product. Therefore search engines may need to develop a taxonomy of search objectives and then develop different search strategies for different objectives to become more successful. A search engine could ask a user whether he is looking for a specific site; e.g. a site which he knows exists and which he might even have visited before. In that case, an appropriate strategy would be to identify and refine key properties of the site. A different strategy would be employed if the user is looking for an introduction into a topic. In that case, a two step search strategy could be used, where first relevant aspects of the topic might be identified. Searches for the aspects would follow. The result would be a reading plan which would allow the user to systematically navigate to sites and keep track of their progress. These strategies might require changes during the indexing phase. Additional properties might need to be recorded for each page.

The holy grail of search engines is natural language understanding. While full understanding is still in the distant future, the partial extraction of meaning may yield significant benefits. Advances in NLP, for example, make it possible to identify the broad topic a text deals with. It can help to distinguish among different meanings of synonymous terms. When the term 'pressure' occurs in a text, for example, the context may be examined to determine whether it is used in the physical, the psychological or one of the other main senses. In a query, the user would be asked to specify the sense; in some cases the system may even be able to extract the intended meaning from the context of the query.

In summary, the current focus of search engines on the identification of words within text is questionable. Approaches are needed which are tuned better to the varying information needs of the user and which capitalize at least partially on the meaning or structure of a document. It may be better, for example, to distinguish 'research papers' from advertising material and indexing only the former word by word while extracting key features such as company, product, context, etc. from the latter.

2.4.5.3. Standards

Standards	Java	JavaScript	XML	Active server pages	Dynamic html	Other

Online knowledge infrastructures differ in the extent to which they support technical standards. Many knowledge infrastructures employ scripting languages such as JavaScript on the client side. In addition, Java is used increasingly on the client as well as on the server side. Microsoft is trying hard to make Active Server Pages a de-facto standard. Similar issues apply for dynamic html. Extended Markup Language (XML) is currently emerging as the next core standard for the WWW. Its key advantage over HTML is the separation of content and presentation.

2.4.5.4. Document authentication

Document authentication	No	Yes

The ease with which online content can be copied may require special protection schemes to ensure the authenticity of a document. Several proposals exist for protecting document authenticity (Mintzer et al. 1998). However, no current scientific knowledge infrastructure implements such mechanisms.

2.4.5.5. Duration of record-keeping

Duration of record-keeping	Permanent	Transient

Similarly to traditional scientific journals, most knowledge infrastructures aim to preserve scientific knowledge for posterity. This goal is more difficult to achieve in an online environment than on paper. While paper documents can survive for centuries without being attended to, online infrastructures require constant maintenance.

2.4.5.6. Support for sessions

Support for sessions	No	Yes

The http protocol underlying the World Wide Web has the disadvantage of being stateless. A request sent to a web server is independent of the previous request; a web server cannot detect that a reader has left its pages or closed his browser. Therefore it is difficult to support interactions consisting of several steps - such as submitting an article, categorizing its keywords, checking consistency, confirming storage location etc. Therefore an increasing number of electronic journals layer custom session mechanisms on top of the HTTP protocol. These approaches are frequently based on 'cookies' - persistent variables which are stored by a browser and which can be retrieved by the web server. Another alternative is to add a unique session variable to all requests.

3. Structure of scientific knowledge

The previous chapter has shown that a wide range of current knowledge infrastructures exists. The infrastructures differ significantly in the way that they represent and manipulate scientific knowledge. Online journals, for example structure knowledge in the form of individual, largely self-contained articles. Encyclopedias, in contrast, focus on the integration of different pieces of knowledge and on their relationships. Moreover, most online journals and encyclopedias treat knowledge as something that can be stored independently of author and reader. However, collaboratories and online forums view knowledge as arising in the context of ill-structured communication processes and as something that is not independent of the individual.

Although these assumptions directly affect the architecture and functionality of an infrastructure, they are rarely evaluated systematically. Design decisions are frequently based on tacit assumptions about the nature of knowledge which later may turn out to be inadequate or inconsistent. As a consequence, it is necessary to analyze the characteristics of scientific knowledge in detail. In the field of information systems, modeling techniques are widely used to systematically describe the structure of a target domain. Such methods will be used to develop a conceptual model of scientific knowledge. Before the model can be formulated, however, the objectives need to be clarified. They consist of facilitating the technical development of infrastructures, of developing epistemologically adequate infrastructures (i.e. infrastructures which correspond to researchers' conceptualizations of scientific knowledge) and for standardizing scientific knowledge. To develop a model, first the general concept of 'knowledge' needs to be examined. This sets the stage for identifying a generic 'unit of scientific knowledge' and for determining its attributes. Finally a conceptual model of scientific knowledge can be derived. It will consist of different categories of knowledge and will include their relationships and properties.

3.1. Objectives

In this section, the objectives of characterizing the structure of scientific knowledge will be determined. Three different types of objectives need to be addressed:

3.1.1. Information systems development

A systematic conceptualization of the structure of scientific knowledge is needed for the development of a knowledge infrastructure. Design decisions depend directly on the structure of the underlying domain. Every information system is explicitly or implicitly based on a conceptualization of the domain in which it operates. During development, analysts and developers translate their assumptions and their knowledge about the domain into source code, table definitions, user interface designs etc. Such knowledge is typically formulated as a conceptual model. Examples for conceptual models are entity-relationship models, object-oriented class diagrams, event-oriented process chains, etc.

Thus an analysis of the structure of scientific knowledge needs to lead to a conceptual model of scientific knowledge. The latter then determines the design of the database which is at the core of a knowledge infrastructure. Reliable techniques exist to transform the elements of a conceptual model - e.g. classes, attributes and associations - into tables and referential integrity relationships of a relational database. Alternatively, such elements can be translated into the structures of an object-oriented database.

Besides database design, the conceptual model is also needed for designing processes and supporting software algorithms for manipulating scientific knowledge. Search algorithms, for example, may depend on type of knowledge or on the overall structure in which a particular unit of knowledge is embedded. A better understanding of the structure of scientific knowledge is also needed for the development of algorithms which identify patterns of knowledge as well as inconsistencies and interdependencies in the overall body of knowledge.

The relationship between the design of an infrastructure and theoretical conceptualizations of scientific knowledge is not a one-way street. Every knowledge infrastructure can be regarded as a test for the adequacy of the underlying conceptualization. Thus the evolution of scientific knowledge infrastructures can also be interpreted as a competition between different theories of knowledge: inadequate theories will be discarded because infrastructures which are based upon them will not survive while useful theories are refined and extended.

3.1.2. Epistemological adequacy

The way in which scientific knowledge is represented in a knowledge infrastructure should closely correspond to the way in which scientists think about scientific knowledge and the way in which they manipulate scientific knowledge. The theory of knowledge in scientists' minds and the theory of knowledge that is implicit in a knowledge infrastructure should match as closely as possible. This objective will be called 'epistemological adequacy (of the infrastructure)'. The term 'epistemological adequacy' was originally introduced as general criterion for assessing the quality of representations: "A representation is called *epistemologically adequate* for a person or machine if it can be used practically to express the facts that one actually has about the aspect of the world." (McCarthy & Hayes 1969, p.27).

This objective increases the acceptance by the scientific community and simplifies the configuration of research processes.

The current, paper-based publishing system does not score well on this criterion. Although researchers have become used to the current structure of articles, references and journals a significant gap exists between the units of scientific knowledge such as theories, methods, concepts, etc. and the way that knowledge is represented in scientific journals. At present it is difficult to answer even basic questions about the body of scientific knowledge. The following statement illustrates the problem:

> "I tried to review all published empirical literature on [***]. Searching for such studies is difficult because they are published in a variety of disciplines." … "Furthermore, the terminology that authors use in [***] varies, so it is difficult to locate relevant papers. The term '[***]' often does not appear in titles, and the titles for many references in this paper would fail to alert the searcher." (Armstrong 1997, p 64). [***] is used to generalize the statement beyond the author's specific topic).

One approach to assess epistemological adequacy is to formulate competency questions, i.e. questions about scientific knowledge that an information system should be able to answer (Uschold & Grüninger 1996). The above statement illustrates one competency question: finding all entries in the knowledge infrastructure that relate to a specific topic or a specific term. This is difficult not only because of the limitations of paper but also because of the limitations of keyword indexing schemes. The current environment has many more epistemological inadequacies (see Table 13). It is not possible, for example, to determine whether knowledge contained in an article is still valid or whether it has been invalidated by subsequent findings (C5). It is not possible to determine which theories are available in a specific area (C13) or which extensions and variants to a theory have been proposed (C6). Other questions relate to the dependencies between articles (C3, C6, C7). References currently only point from the citing to the cited article; however it would often be instructive to know which other articles build on a paper (C7). Table 13 lists several additional competency questions.

These questions show significant limitations of the current body of knowledge. Only very few questions – e.g. regarding the author – can be answered. The problem can be traced to the current explicit structure of scientific knowledge which had been shown in Figure 3 on page 13 as an object-oriented class diagram. Only questions which relate to classes in the diagram can be answered. Thus an article's authors can be identified, as well as its descriptors and the articles which it references. Some additional difficulties arise from the lack of integration of knowledge: currently it is not straightforward, for example, to identify all articles which have been authored by a person. Similarly, it is not possible to analyze references in the opposite direction (e.g. to answer the question which other contributions have built on an article).

The source of these problems is obvious: the structure of the traditional publishing system is not based on epistemological considerations; it is driven by the needs and capabilities of paper. Thus, the current body of knowledge is structured around concepts which are meaningful for the processes of editing, printing and distribution, yet epistemologically meaningless: the basic unit of current knowledge is the article. It is merely a container for any kind of scientific or non-scientific ideas.

Table 13. Competency questions for epistemological adequacy and the capabilities of the traditional paper-based scientific publishing system

No	Competency question		Comment
C1	Who wrote this article?	+	Supported (author)
C2	What are the article's main points?	+	Supported (abstract)
C3	On what previous research is this article based?	p	Partially supported via references, but many implicit dependencies are not shown
C4	What are other contributions by the same author?	-	Not supported except via citation index database
C5	Is the knowledge in this article still valid?	-	Not supported
C6	What extensions have been proposed?	-	Not supported
C7	Have other articles referenced this article? Which?	-	Not supported except via citation index database
C8	Who is working on the same problem?	p	Very limited support via references
C9	What solutions have been proposed for problem X	-	Not supported
C10	What articles address issue X?	-	Not supported
C11	What are unsolved problems in domain X?	-	Not supported
C12	What topics lack research activity in domain X?	-	Not supported
C13	What theories are available in domain X?	-	Not supported
C14	Which theories incorporate concept X?	-	Not supported
C15	What are alternative interpretations of concept X?	-	Not supported (but may be described in article)
C16	What are the most important contributions?	p	Partially supported via journal rankings
C17	What is the quality of article X?	p	Partially supported via journal rankings

+) supported; p) partially supported; -) not supported

In contrast, an epistemologically more adequate body of scientific knowledge requires a structure which reflects the core concepts distinguished in the theory of knowledge and in the practice of science such as 'theory', 'concept', 'data' etc. Certainly, these elements are not absent from the current body of knowledge. They appear in many articles. But they are not structurally embedded in the current infrastructure. Therefore they cannot be targeted by automated or semi-automated processes and procedures. Many of the competency questions described above could be answered automatically, if the epistemological elements in each articles were identified and marked as such.

Epistemological adequacy is the key criterion for the development of a knowledge infrastructure. To increase the epistemological accuracy of a knowledge infrastructure, therefore, it is necessary to analyze the structure of scientific knowledge in detail and to determine the elements of scientific knowledge with their properties and relationships.

3.1.3. Standardization

Scientists and philosophers have long sought to establish standards for scientific knowledge to simplify collaboration and the exchange of ideas. Although much has been achieved, a significant gap remains between philosophical conceptualizations of scientific knowledge and scientific practice. The goal of a cumulative body of knowledge requires a minimum agreement about the nature of scientific knowledge. The stronger the integration of knowledge, the more important it becomes to agree on a standard about the elements of scientific knowledge. By identifying core elements of scientific knowledge such as theories, variables, problems, methods etc. and by clarifying their relationships a conceptual model of scientific knowledge can provide a foundation from which such standards can be developed.

Because science is highly decentralized and highly collaborative, establishing a common understanding of the characteristics of knowledge may greatly increase research productivity. A common conceptual framework reduces the effort for mapping a partner's results into one's own scientific schema. It improves the integration of knowledge as well as quality control. In addition it may be useful as a knowledge interchange format. An explicit conceptual model can be used to exchange knowledge between scientific knowledge infrastructures which use different approaches for storage. By identifying the key attributes and relationships between elements of scientific knowledge, a conceptual model provides an intermediate structure to which knowledge from different infrastructures can be mapped.

3.2. Theoretical foundations

3.2.1. Concept of scientific knowledge

The nature of knowledge has been a matter of intense discussion since the beginnings of philosophy. The philosophical discipline which studies the concept is called 'epistemology'. The term is a combination of the Greek words 'episteme' (knowledge) and 'logos' (theory) (Bierman & Gould 1981, p 141). Key problems of epistemology are clarifying the characteristics of knowledge and distinguishing knowledge from other concepts.

Much research has sought to refine the concept of knowledge and to answer questions about its core characteristics. However, a universally accepted interpretation has not yet emerged. Therefore the term knowledge must initially be treated as a sociological phenomenon:

> "Every theory of knowledge must start from knowledge as a given sociological fact. The system of knowledge as it has been built up by generations of thinkers, the methods of acquiring knowledge used in former times or used in our day, the aims of knowledge as they are expressed by the procedure of scientific inquiry, the language in which knowledge is expressed - all are given to us in the same way as any other sociological fact such as social customs or religious habits or political institutions. The basis available for the philosopher does not differ from the basis of the sociologist or psychologist; this follows from the fact that if knowledge were not incorporated in books and speeches and human actions, we could never know it. Knowledge therefore is a very concrete thing; and the examination into its properties means studying the factors of a sociological phenomenon." (Reichenbach 1938, p 3)

The most common approach to studying the concept, therefore, is to treat the concept as undefined and to approximate its meaning by examining the context of its use. An analysis of the use of the term in everyday and scientific language leads to three main interpretations which will be discussed in the next sections: knowledge-that (objective knowledge), knowledge how (know-how) and knowledge by acquaintance.

3.2.1.1. Knowledge-that

The term 'knowledge' is often used in phrases which indicate that someone 'knows' something. These sentences typically consist of a proposition which is made by a person and which is either true or false. The fact that the person 'knows' the proposition is an indication for the reliability of the statement. A person who knows something indicates that he has a higher confidence in the truth of the proposition than if other terms such as 'believe' or 'hope' were used. Knowledge-that traditionally is associated with the following requirements:

1. It must contain a proposition which can be either true of false.
2. There must be a person whom the knowledge is ascribed to.
3. The holder of knowledge must be able to justify the proposition.
4. The holder of knowledge must believe the knowledge.
5. The proposition must be true.

This interpretation of knowledge is frequently described as 'justified, true belief' (Dancy 1985, p 23). It implies that a correct prediction of a chance event cannot be considered knowledge. It also has important practical advantages: It identifies a basic unit of knowledge: a statement or proposition. It treats knowledge as explicit; propositions and justifications always need to be formulated in some language and therefore can be separated from mind. Thus this type of knowledge can be made available in a computer system. This interpretation is closely related to logic – the science of reasoning. Techniques of logic can be utilized to decide whether a proposition that depends on other statements (facts, other propositions) is true or false. Unfortunately, several fundamental problems exist which significantly reduce the attractiveness of logic-based 'knowledge-that' for building an integrated body of scientific knowledge. They will be discussed in section 3.2.2. below.

If this conceptualization of knowledge were used for the development of a knowledge infrastructure, then knowledge would be represented as collections of formal statements. Researchers would add knowledge by creating facts and propositions and by assigning truth values to each statement. Logic algorithms would be used intensively to prove or reject propositions. Conceptually, the knowledge infrastructure would consist of four components: a set of facts – which are always true –, a set of rejected 'false' propositions, a set of propositions which have not yet been accepted or rejected and a set of propositions which have been found to be true. The main criterion for scientific progress would be the increase in the number of true propositions.

The interpretation of knowledge as knowledge-that is attractive because it leads to identifiable components of knowledge. This perspective provides a basis for investigating the structure of knowledge and the ways in which its elements interrelate. Unfortunately, the limitations of this conceptualization become apparent when it is applied in practice: knowledge is often highly complex and interrelated with many other elements of knowledge that are difficult to formulate in clear, precise statements. This leads to a different conceptualization of scientific knowledge which avoids any hypothesis about structure and components of knowledge but which attempts to explain knowledge from its impact: knowledge-how.

3.2.1.2. Knowledge-how

The term 'knowledge' is often used in phrases which describe how some result has been achieved. The fact that a person is capable of riding a bicycle or flying an airplane, for example, can be attributed to her knowledge. This interpretation focuses on the factual implications of knowledge; it does not require assumptions about components of knowledge. The concept of knowledge-how arises from an outcome perspective which treats the human mind mostly as a black box. Knowledge-how is something which enables a person to perform well in certain situations. This type of knowledge can be identified by examining the situations in which knowledge is evident and by contrasting it from other factors on which performance depends. This includes physical factors such as strength and endurance as well as mental factors such as motivation, creativity and intelligence which are typically regarded as being different from knowledge.

Knowledge-how and knowledge-that are not equivalent. Some knowledge-how cannot be described in the form of knowledge-that. A cyclist, for example, has knowledge-how. However, he is usually at loss when asked to explain how to ride a bicycle. In other cases, knowledge-how and knowledge-that may overlap: someone may know how to perform a task and be able to communicate that knowledge.

Besides the distinction between knowledge-that and knowledge-how, there is also another distinction between explicit and tacit knowledge. Both pairs are not equivalent. Tacit knowledge is knowledge about which a person is only partially aware and which he cannot communicate. Explicit knowledge is the opposite. Thus tacit and explicit knowledge are mutually exclusive. Both pairs of concepts differ in the frame of reference. Knowledge-that and knowledge-how are based on criteria which are external to the person carrying the knowledge while tacit and explicit knowledge are based on internal criteria.

A problem of 'knowledge-how' is the inability to identify a unit of knowledge. As performance is used as evidence for the presence of knowledge-how in certain activities, a performance unit might be used. However, because of the variety of possible activities this would lead to considerable aggregation problems. While it is possible to determine activities which require knowledge it is difficult to compare knowledge-how across activities.

3.2.1.3. Knowledge by acquaintance

A third perspective focuses on the origins of knowledge. It is closely related to the question "what can we know?" Russell pointed out that a form of knowledge needs to be distinguished which does not need any further justification: This is the knowledge which we pick up with our senses. Witnessing an event or 'knowing people' are examples of knowledge by acquaintance. Russell distinguished knowledge by acquaintance from knowledge by description and argued that all knowledge ultimately needs to be based on knowledge by acquaintance (Russel 1918). Historical knowledge such as the fact that Caesar lived in ancient Rome, is knowledge by description. Although this knowledge has been passed on through the centuries, it can be traced to persons who had 'firsthand' knowledge of Caesar – who knew him by acquaintance. However, the idea that our senses report an accurate or objective view of the world has been widely discredited. Nevertheless, the distinction between different sources of knowledge and especially the distinction between knowledge which follows from other knowledge and axiomatic knowledge remains important and needs to be supported by a knowledge management infrastructure.

3.2.2. Problems related to knowledge

Many efforts for the development of knowledge infrastructures explicitly or implicitly strive for the goal of creating an objective, integrated, consistent and true body of scientific knowledge. Design decisions depend on this vision; algorithms and mechanisms are often developed to come closer to this goal. This is particularly visible in logic servers which use formalisms of predicate logic as the foundation of their knowledge infrastructure. However, many fundamental problems have been identified in epistemology which limit the ability to create an explicit and consistent body of scientific knowledge. These problems include the existence of reality, the reliability of sense-impression, the foundation of concepts, as well as the existence of absolute truth. Some of these problems have direct implications for the development of an infrastructure of scientific knowledge and therefore will be analyzed in the following subsections:

3.2.2.1. Existence of absolute truth

The ideal of a large body of true statements which reflects the achievement of the sciences is attractive, but it encounters a number of practical and fundamental problems. The first is the problem of absolute truth; i.e. the difficulty of knowing with absolute certainty that something is true. Many philosophical objections exist. An extreme position denies certainty of any sense-experience. We cannot be sure that our senses portray a correct image of the world or that the world is not a dream. While Descartes concluded that we can be sure about our own existence, it is doubtful what else we can be certain of. A second issue is the existence of error. Humans are used to making errors which are only recognized later. How can one be sure that a seemingly perfect contribution in a knowledge infrastructure is devoid of error? Another objection is the problem of induction which was raised by Hume (1748): What is true in the present need not remain true in the future. Thus observations can never establish the absolute truth of a theory.

The larger a collection of propositions, the more problematic the assumption of absolute truth becomes. There seems to be agreement that the ideal of an absolute truth is not possible; therefore absolute knowledge is not possible. Kuhn's historical investigation into scientific theories supports this view by showing that widely accepted theories often contain elements which are known to be inconsistent with observation and therefore false from a logical perspective (Kuhn 1996). In practice, scientific knowledge can be seen as constructed by the scientific community. Researchers as a group give credit to theories not just through their beliefs, but also through their actions (Chalmers 1990), by incorporating them into "their ongoing process of research production." (Knorr-Cretina 1981, p 8)

The inability to reach absolute truth has significant implications for the design of knowledge infrastructures:

- Any knowledge infrastructure which sharply distinguishes between 'true' and 'false' knowledge will not be able to represent a large part of the current body of scientific knowledge. Knowledge infrastructures cannot be based on the concept of absolute truth. They must be able to support conflicting assessments of truth. In addition, they may need to support different degrees of truth (see the section on modal reasoning below).
- A knowledge infrastructure must support the dialogue about the correctness of knowledge. It must be capable of supporting dramatic changes in the evaluation of knowledge which may result from paradigm changes when new foundational theories replace theories which previously have been held in great esteem and as a consequence a large portion of the body of knowledge has to be reevaluated.

3.2.2.2. Hermeneutics – the problem of meaning

A second problem associated with the traditional conceptualization of knowledge derives from the difficulties associated with the meaning of a linguistic concept. In the same way as one cannot take the concept of knowledge for granted, concepts such as 'temperature', 'atoms' etc. are open to interpretation and their meaning may change over time. The philosophical stream concerned with such problems is called hermeneutics (Gadamer 1976). It has its roots in the translation of ancient religious texts and the problem of finding the 'right' translation. The latter was eventually determined to be an impossible goal (Gadamer 1976). Today the term hermeneutics has been extended to the universal problem of sense-making and the assignment of meaning which humans continuously face (Boland 1994).

The problem of interpretation also arises for scientific concepts. Such concepts are ideally defined with the help of other, already established scientific concepts. However, to avoid an infinite regress, every body of knowledge must contain some foundational concepts which are not defined in this way. The semantics of these concepts cannot be taken for granted. Definitions of foundational concepts make use of non-scientific and ambiguous language which often include references to concepts which are defined using these foundational concepts. Circularity in definitions is the consequence and circularity cannot be avoided. Another problem is that interpretations of concepts may change over time. As Kuhn has pointed out, even seemingly stable concepts such as 'space and time' have undergone significant change in meaning (Kuhn 1996, p 102).

As a consequence, the following two issues arise for the development of any infrastructure dedicated to knowledge:

- It is not possible to endow the body of knowledge with precise meaning. While Gadamer (1976) had initially regarded this as true only in the humanities, a similar problem arises in the natural sciences (Connolly & Keutner 1988, p 36). As a consequence, formal systems of knowledge can only be built on "shaky foundations" (Popper 1968, p 111). Variations in the interpretation of foundational concepts cannot be avoided. Knowledge infrastructures must be able to distinguish different interpretations of a concept and they must be able to track the change in meaning over time.
- A knowledge infrastructure must provide mechanisms for the process of interpretation. This is a subjective process which needs personal (positive) prejudices and self-reflection (Gadamer 1976). A knowledge-infrastructure therefore must provide mechanisms which allow different readers to access and interpret the same source in different, personal ways and which should support the reader in integrating the knowledge into their personal knowledge horizon.

3.2.2.3. Cumulative nature of knowledge

Another problem for knowledge infrastructures is the way in which scientific knowledge accumulates. Ideally, the body of knowledge grows gradually as new insights are added and incorrect theories are removed. For the technical elements of an infrastructure this would mean that a regular stream of additions and removals of scientific knowledge could be expected. However, Kuhn has argued forcefully against the idea of science as the linear accumulation of knowledge. He distinguishes two phases of science: Normal science and scientific revolutions.

> "Normal science, the activity in which most scientists inevitably spend almost all the time, is predicated on the assumption that the scientific community

> knows what the world is like."...[It] "often suppresses fundamental novelties because they are necessarily subversive to its basic commitments." (Kuhn 1996, p 5)

Kuhn attributes a central role in the research process to the "constellation of beliefs, values, techniques, and so on shared by the members of a given community" (Kuhn 1996, p 175) which he calls a "paradigm". He describes normal science as 'puzzle-solving':

> [Normal science]... "seems an attempt to force nature into the preformed and relatively inflexible box that the paradigm supplies. No part of the aim of normal science is to call forth new sorts of phenomena; indeed those that will not fit the box are often not seen at all. Nor do scientist normally aim to invent new theories, and they are often intolerant of those invented by others. Instead, normal-scientific research is directed to the articulation of those phenomena and theories that the paradigm already supplies." (Kuhn 1996, p 24)

While normal science gradually increases the amount of available knowledge, from time to time events occur in which the whole belief system of the scientific community is shattered and replaced with a new one:

> [Only when]... "the profession can no longer evade anomalies that subvert the existing tradition of scientific practice – then begin the extraordinary investigations that lead the profession at last to a new set of commitments, a new basis for the practice of science. The extraordinary episodes in which that shift of professional commitments occurs are the ones known" ... "as scientific revolutions." (Kuhn 1996, p 6)

Kuhn's account of science raises concerns for the design of knowledge infrastructures. While knowledge infrastructures can support the growth of knowledge in linear fashion it is not clear how revolutions which replace and reinterpret much of the previous body of knowledge can be handled. For example, if the infrastructure consisted of separate areas for accepted knowledge which fit into the current paradigm, for uncertain knowledge that is neither accepted nor rejected and for clearly rejected knowledge, then a paradigm shift might suddenly eliminate most of content of the infrastructures' core area for accepted knowledge. On the other hand, if an infrastructure were based on the interpretive paradigm, it would not contain sharp boundaries between accepted and rejected knowledge. It would allow different, possibly inconsistent interpretations to co-exist at the same time. This would allow for the co-existence of different paradigms and therefore for a smoother transition.

This leads to the following consequence for the design of knowledge infrastructures:

- Its structure must be independent from a specific scientific paradigm. It should allow different paradigms to coexist and must not enforce sharp boundaries between accepted and other types of knowledge.

3.2.2.4. Demarcation of scientific knowledge

Many philosophers of science have attempted to identify a criterion which distinguishes 'scientific knowledge' from 'common-sense' and other forms of knowledge (Popper 1968, p 18). Such a criterion would be highly important for the creation of scientific knowledge management infrastructures as it would have an impact on the way that such knowledge needs to be represented and manipulated. If the truthfulness of a scientific idea were such a criterion then a knowledge infrastructure would need to be split into different sections. One section would contain all true ideas, while other sections might contain ideas which are under investigation; a third section might preserve ideas which have been proven to be false. The most valuable part of such an infrastructure would be the section with true ideas. Services and functionality provided by the infrastructure would differ from section to section. However, as Popper (1968) has shown – the truth of a scientific idea can never be proven; it can only be falsified. The latter criterion, in turn has been questioned by other authors such as Kuhn (1996). It follows, that scientific knowledge infrastructures cannot adopt a strict separation between scientific and non-scientific knowledge.

3.2.2.5. Structure of categories

Category systems are a fundamental component of science. To solve a problem by using scientific knowledge, it is necessary to map real-world phenomena to the corresponding categories. Much thought has been spent over the millennia on the nature of such categories. Aristotle established the classical distinction between optional, necessary and sufficient properties. Necessary features define the 'essence' of a thing. In addition, his properties are binary: they either apply or don't apply to a thing. It follows that "features have clear boundaries.." and that "..all members of a category have equal status." (Taylor 1995, pp 23-24)

This classical position is still predominant in many areas of science. First order logic, the most widely used system of logic, is based on discrete categories. Knowledge representation languages such as KIF are also based on discrete categories. Object-oriented development methodologies attempt to structure a domain into hierarchies of classes with clearly defined boundaries (Booch, Jacobson & Rumbaugh 1998).

There has been substantial evidence, however, that human linguistic categories do not exhibit these properties. Wittgenstein illustrated the problems using the category 'Game'. He pointed out that it is not possible to identify features which apply to all types of games:

> "Consider, for example the proceedings that we call 'games'. I mean board-games, card-games, ball-games, Olympic games, and so on. What is common to them all? - Don't say: 'There must be something in common, or they would not be called "games"' - but look and see whether there is something that is common to all. - For if you look at them you will not see something that is common to all, but similarities, relationships, and a whole series of them at that. To repeat, don't think, but look! - For example at board games, with their multifarious relationships. Now pass to card-games; here you find many correspondences with the first group, but many common features drop out, and others appear. When we pass next to ball-games, much that is common is retained, but much is lost. - Are they all 'amusing'? Compare chess with noughts and crosses. Or is there always winning and losing; but when a child throws his ball at the wall and catches it again, this feature has disappeared. Look at the parts played by skill and luck;

> and at the difference between skill in chess and skill in tennis. Think now of games like ring-a-ring-a-roses; here is the element of amusement, but how many other characteristic features have disappeared! And we can go through the many, many other groups of games in the same way; we see how similarities crop up and disappear.
>
> And the results of this examination is: we see a complicated network of similarities overlapping and criss-crossing: sometimes overall similarities, sometimes similarities of detail.
>
> I can think of no better expression to characterise these similarities than 'family resemblances'"... "And I shall say 'games form a family'.
>
> How is the concept of a game bounded? What still counts as a game and what no longer does? Can you give the boundary? No. You can draw one; for none so far has been drawn. (But that never troubled you before when you used the word 'game'.)" (Wittgenstein 1978, pp 31-33)

In independent experiments, Labov and Rosch studied the boundaries between linguistic categories. Labov showed subjects line drawings of household receptacles such as cups, mugs, bowls etc. and asked them to name the objects (Labov 1973). He could show that there was no clear boundary between the concepts. Instead, a gradual transition between categories was found. Rosch (1973) asked 200 American college students "to judge to which extent each of sixty household items could be regarded as a good example of the category furniture." (Taylor 1995, p 43) On the top of the list were the items 'chair', 'sofa', 'couch'; on the bottom: 'ashtray', 'fan', 'telephone'. The order was statistically highly reliable.

Both researchers concluded that categories do not require clear boundaries. Instead, categories tend to be organized around good, clear exemplars, which are termed 'prototypes' (Taylor 1995, p 43). The more similar an item is to a prototype, the more likely it is to be identified as a member of the category. In many additional experiments, the existence of prototype categories have been confirmed (Taylor 1995, p 48). Additional research has focused on the relationship between categories. Categories can be organized in hierarchies where each subordinate term is a specialization of the superordinate term. In the classical view, within a hierarchy, only the most general and the most specific should have special status. Experiments, however, show that there exists "a level of categorization which is cognitively and linguistically more salient than others"...:

> "If a foreigner were to point to the object I am not sitting on and ask 'What do you call that in English?', I would almost certainly answer, 'It's a chair.' I would not reply 'It's an artifact', or 'It's a piece of furniture' even though these alternative answers would be equally 'correct'."(Taylor 1995, p 49)

The most salient level of categories is frequently called the basic level (Taylor 1995, p 50).

This has important implications for the design of electronic knowledge infrastructures which aim to support a substantial set of human linguistic categories:

- Categories are not binary. It must be possible to distinguish different degrees of category membership. Some members may be central to the category, while others may be at its fringes. The system should be able to distinguish between more and less central exemplars of a category. It must be able to identify the degree of membership in a category. For any exemplar, it must be able to provide an ordered list of categories to which it belongs.

- A category system must support overlapping categories where one item may fit into several categories with different degrees.
- A category system must be able to distinguish more salient, basic-level categories.
- The system may have to include some rules regarding the cut-off for membership in a category.

3.2.2.6. Complexity

Several additional problems of conceptualizing knowledge are rooted in practical and fundamental problems of reasoning. In the following, three such problems will be discussed: the problem of complexity, and the problems of nonmonotonic and modal reasoning.

Science has been very successful in breaking large problems down into smaller, more manageable pieces, which can be analyzed separately. This analytic approach sometimes disregards interdependencies between a system's elements which may be crucial for its behavior as well as for the behavior of its parts. Systems which are difficult to break down into small, independent components are called 'complex'. Such systems typically consist of a large number of elements with dynamic and non-linear interaction; they are open systems which do not operate in an equilibrium and their state depends on their history. Examples are the human brain which consists of trillions of neurons or the national economy which consists of millions of actors (Cilliers 1998, p 3f).

A measure for complexity is the shortest possible description of a system (Chaitin 1975; Casti, 1995, p 9; Shannon & Weaver 1949). It implies that the representation of more complex systems may significantly increase the information processing requirements for a knowledge infrastructure. The more complex a system is the more necessary it may become to control the information processing load on the knowledge infrastructure and to develop custom information processing solutions. When it is impossible to create complete models of a complex system, the study turns to heuristics and partial solutions which often become an object of research in their own right. Whenever a researcher encounters systems which are more complex than his information processing capabilities can handle, he must choose a strategy which optimizes the tradeoff between error and cognitive load. Therefore researchers frequently use greatly simplified models with the expectation that their error is acceptable in the vast majority of cases. Thus the study of complex systems may at times more resemble an art where researchers seem to follow fashions and the rationale for certain behavior may be elusive. However, knowledge infrastructures should not penalize such behavior because this would risk leading researchers to study less complex but also less significant problems.

A second aspect of complexity are the fundamental limitations in the cognitive capabilities of humans which need to be taken into account during the study of social systems. The representations which humans form of their environment are necessarily inaccurate and limited. Even if the world were fully deterministic, once humans become part of the picture, researchers need not only formulate an accurate model of the 'real world'. They also need to create an accurate model of the skewed and simplified representation humans make of this world. Thus intelligent beings significantly increase the complexity of scientific models.

This leads to the following implications for knowledge infrastructures:

- Any system which captures and organizes complex knowledge may exhibit significant inconsistencies. In a highly complex world where human decisions depend upon incorrect and incomplete representations of their environment which may be difficult to identify, incompatible theories may need to coexist for extended lengths of times. A knowledge management system for the social sciences therefore cannot require that its contents must be consistent. Otherwise, a substantial body of research knowledge would have to be excluded.
- Mechanisms are needed to measure the complexity of the underlying phenomenon.

3.2.2.7. Nonmonotonic reasoning

Knowledge infrastructures which are based on traditional logic have an important limitation: they require that reasoning processes are monotonic. A logical system is said to be monotonic if a conclusion which can be logically proven from a subset of all true facts always remains true when all facts contained in the knowledge base are taken into account. Human reasoning, however, is frequently not monotonic. In "real-world" situations, additional facts often change the conclusion that is drawn (McCarthy 1990). An example is a rule for the selection of a transportation medium: "If the distance is larger than 250 miles, then travel by plane." This rule can be used to represent the knowledge that a person would decide to take the plane for traveling from Los Angeles to San Francisco. However, this rule would no longer be accurate if the additional fact were discovered that the airlines are on strike. In this case, the traveler would alter his decision based on the additional fact. Of course, the rule in the knowledge base could be modified to address this case: "If the distance is larger than 250 miles and the airlines are not on strike, then travel by plane". However, this approach has the disadvantage of leading to very complex rules because all possible exceptions (e.g. that the airport is closed, the flight has been canceled, etc.) need to be explicitly included. This also implies that the decision-maker needs to verify that none of the potential exceptions has occurred. Thus even simple problems may quickly become intractable. Although much progress has been made in the development of nonmonotonic logics it cannot yet be applied in situations of realistic complexity. As a consequence, knowledge infrastructures cannot be based on the assumption that their content allows monotonic reasoning. In addition, infrastructures need to provide support for automatic or interactive nonmonotonic reasoning algorithms.

3.2.2.8. Modal reasoning

Traditional logic is based on two discrete truth values: true and false. These two values are not always sufficient. Frequently, it is not necessary to distinguish whether a statement is true but whether it could be true. In other cases, humans need to assess whether a statement is desirable. These modalities such as possibility, necessity, usefulness, can be introduced as predicates into traditional logic. However, many questions about the semantics of these modal operators remain unsolved. As a consequence, the formalization of statements which involve possibility, necessity and usefulness still entails significant problems. The social sciences suffer more than the natural sciences from the absence of practically applicable systems of modal logic. As a consequence, knowledge infrastructures which include knowledge about human behavior, need to include mechanisms for handling additional modalities.

3.2.3. Relationship with paradigms of science

Scientists do not agree about 'science'. They differ in their fundamental assumptions about the world, in the assessment of the objectives of research, in their interpretation of the formation of concepts and many other issues. The differences have considerable impact on the selection of research problems, on the methods used for conducting research and on the evaluation of research results. Researchers who – knowingly or unknowingly – follow different philosophical paradigms work differently and therefore may need different kinds of support from a knowledge management infrastructure.

The differences between these philosophies need to be analyzed to determine their impact on the development of knowledge management infrastructures. It may be that the philosophies are so fundamentally different that an integrated infrastructure is not possible. Another outcome of the analysis may be that all philosophies share common elements on which the core of a knowledge management infrastructure can then be built. The analysis will first focus on the three main paradigms in the philosophy of science: positivism, the interpretive/constructivist paradigm and the critical paradigm. The suitability of each as a foundation for knowledge management infrastructures will be evaluated.

Table 14 lists the main characteristics of the three predominant paradigms (for alternative classifications see Schwemmer (1981, p 56f) or Mitev (2000). The predominant paradigm in social science research including information systems continues to be [logical] positivism which was formulated in the 1920s in Vienna. It is an extension of the work of Auguste Comte who defined positivism in the 19th century. The objective of logical positivism is to

> "..distinguish [science] from metaphysical and religious discourse" ... [and] ... "to construct a general definition or characterization of science, including the methods appropriate for its construction and the criteria to be appealed on its appraisal." (Chalmers 1990, p 3)

Positivism sought to define a universal method of science which is independent of the domain and historical context in which science is conducted. Positivism is based on an empiricist perspective: it places the scientist in the position of an independent observer of an objectively existing reality: there is only one reality which can be uncovered by unbiased, methodical observation. Scientific knowl-

Table 14. Comparison of the positivist, interpretivist and critical paradigms (Peile (1994), p.63 (rows 1 to 6)).

	Positivism	Interpretivism /Constructivism	Critical paradigm
Cosmological assumption	Causal determinist view of reality. The world is predictable, knowable and measurable. Fragmentary view of reality (reality can be understood as separate parts)	Knowledge is contextual and a symbolic social construction. Events can be explained and their meaning for people uncovered. Parts can only be understood in context.	All things are internally contradictory and are in a constant process of movement where all processes form a totality in which each process overdetermines every other.
Ontological assumption	Behaviour can be explained in causal, deterministic ways. It has a mechanistic quality. People are manipulatable and controllable.	Behaviour is intentional and creative. It can be explained but is not predictable. People shape their own reality.	Human behavior is social and historic. People shape their world but are shaped by it at the same time.
Epistemological assumption	Knowledge arises from experimentation and observation and is grounded in the certainty of sense experience.	Knowledge arises from interpretation and insight and is grounded by empathetic communication with the subjects of research. Symbols, meanings and hidden factors are essential for understanding.	Knowledge arises through action and is grounded in the success of self-conscious action. Research goes beyond appearances to what is essential.
Ethical assumption	A separation between knowledge and values. Science produces knowledge. How it is used is a value, ethical or moral question and is outside the concern of science.	Values are the subject of research. Moral or ethical relativism. Leads to disinterest in ethical issues or anarchistic individualism.	Knowledge and value can not be separated. Committed to happiness and the emancipation of people from oppression.
Spiritual assumption	Rejection of spiritual explanation or a clear separation between science and religion.	Relativism of spiritual beliefs. Such beliefs are important in the social construction of meaning.	The materialistic rejects spiritual beliefs but they are compatible with the idealistic critical approach.
Political assumption	The relationship aim between science and society is control. The value-free stance implicitly supports domination by the established order. Mutually supportive with both high technology capitalism and centralised industrial socialism.	The relationship aim is empathetic communication. Implicitly conservative since there is no structural or historical analysis of society. Mutually supportive with a liberal society allowing individual freedom and self-determination.	The relationship aim is enlightenment. Explicitly change focussed, seeking to challenge the present capitalist system. Supports a socialist or communist environment.
Consequences for the design of knowledge infrastructures	Repository of objective knowledge. Requires mechanisms for the detection and removal of inconsistencies. Highly formalized structure, ideally in first order logic. Emphasis on truth, integration and consistency.	Sea of thoughts of different degrees of imperfection. Emphasis on interpretation, context, and increasing discourse among researchers.	Repository of context-dependent, valuable knowledge. Requires mechanisms for identifying, prioritizing and allocating research problems. Emphasis on values, action and organizing researcher's work.

edge is absolute and consists of theories – systems of general statements which can be shown to be true. Positivism focuses on mathematical formulations of theories, quantifiable observations and statistical analysis (Remenyi et al. 1998, p 33). Positivism often uses the natural sciences, particularly physics, as a model (Chalmers 1990, p 14). The fundamental assumptions of positivism have been widely discredited: Observation – also in research – depends on preconceived notions about the world. Kuhn used Gallileos discovery of the pendulum as an example: while many people trained in Aristetolian physics had observed the same phenomenon before him, they saw a body that was constrained from achieving its natural resting state – a "body that was simply falling with difficulty." (Kuhn 1996, p 119) They saw nothing mysterious or interesting. In contrast, Gallileo observed a body which "almost succeeded in repeating the same motion over and over again, ad infinitum." (Kuhn 1996, p 119) This allowed him to observe properties such as the pendulum's period which his contemporaries would not 'see'.

Another problem is that physics is not an appropriate model for the social sciences because it focuses on inanimate objects. In physics, situations are easier to describe because they don't include the complex cognitive factors which are always present in the social sciences. In addition, in physics, research findings have no impact on the actual behavior of the objects studied. In physics, an observation may be influenced by an experiment. But the discovery of a lawlike regularity has no effect on how nature unfolds. In the social sciences, in contrast, research findings frequently motivate individuals or groups to change their behavior – thereby potentially invalidating the findings in the future.

Interpretivism is the leading alternative to positivism. It is based on the assumption that an understanding of reality is an act of interpretation. Its outcomes depend not only on the external world but also on the interpreter. The consequence is relativism – scientists need to accept that multiple inconsistent perspectives on an issue may coexist and that it may not be possible to decide which is more correct. Interpretation inevitably incorporates part of the interpreter's value system. In contrast to positivism, interpretivism recognizes that research cannot be value-free (Bunge 1998a, p 210) and requires researchers to bring their value system into the open. Interpretivism does not negate the usefulness of positivist methods for certain problems (Heinemann Pieper 1988, p 535) but argues that reality is too complex in the social sciences. A variant of the interpretivist paradigm is constructivism which assumes that social reality arises from the construction of meaning by its participants.

A third alternative is the critical research paradigm which focuses on emancipation. It's currently most recognized advocate is Habermas (1968, 1981). Originally, a critical theory was seen as "any theory which provides a scientific analysis of a particular social situation that accounts for the oppression of a specific group of people, and how that oppression can be overcome." (Fay 1987, p 29) This interpretation excludes many topics in the social sciences and therefore limits the scope of the paradigm. Peile suggested that Hegel's dialectical method must be regarded as the foundation of the paradigm (Peile 1994, p 29). He defines critical research as

> "...an approach that seeks to reveal the dialectical movement of all things (including research) where this movement leads to the demise or negation of the thing 'as it is' and then to the unfolding or establishment of the thing's potential." (Peile 1994, p 30)

The dialectic process is based on three main elements: contradiction, movement and totality. Everything contains its opposite or negation (Peile 1994, p 31). A thing is made up of what that thing is not; what it repels as its opposite (Marcuse 1968, p 123). Contradiction is also manifest in the gap between actuality and potentiality. Everything strives to realize its potential – thereby changing its actuality. The result is movement. Things are not stable and in equilibrium but transient. As a consequence, reality is always changing and its analysis must include a historical perspective. Finally, a thing cannot be separated from its environment: The essence of a thing can only be understood in the totality of all ... [its] relationships." (Peile 1994, p 33)

In practice, the boundaries between the three paradigms are not always sharp. Pragmatic positivism recognizes many of the issues which have been raised by proponents of interpretivism and the critical approach. Interpretivists recognize that standards are needed to evaluate research approaches, but reject the idea that these standards can be universal. Habermas argues that the critical paradigm involves a synthesis of the other two (Bernstein 1985, p 10).

Besides the three paradigms, many other paradigms such as existentialist and postmodern paradigms have been discussed: Peile proposed an 'ecological' and a 'creative' paradigm (Peile (1994).

The choice of paradigm has direct implications for the development of knowledge management infrastructures. A positivist knowledge management infrastructure can treat knowledge as objective and true unless disproven. It can treat knowledge as being independent of author and reader. Developers of a positivist body of knowledge will seek to develop mechanisms for the automatic detection and removal of inconsistencies within the body of knowledge. The positivist's ideal would be a large body of explicit knowledge which would contain the collective knowledge of all scientists and which would be complete and consistent.

Knowledge infrastructures based on the other paradigms require different structures. In both cases the objective of a consistent body of knowledge must be dropped. Mechanisms must be developed to manage incompatible knowledge. In a critical environment, for example, authors may be required to describe the negation of their findings. In both environments, it would be much more difficult to localize knowledge. Knowledge arises through interpretation or action of individuals. Explicit documents only capture part of this knowledge. Therefore interpretive and critical infrastructures require additional effort for clarifying the context and for adding interpretation aids. In addition, a critical infrastructure will seek to identify power structures and add mechanisms for supporting the incorporation of value judgements into the research process. Among all three paradigms, the interpretive is least constraining. It is capable of representing knowledge formulated according to the positivist paradigm and can easily be extended to represent knowledge formulated according to the critical paradigm. Therefore it can be used as the core of a universal knowledge infrastructure which caters to scientists following any of the three main paradigms. Note that adopting a standard representation of knowledge does not imply that an infrastructure is used in the same way for each paradigm. Positivist, interpretive and critical researchers follow different research approaches and therefore require different support functionalities. A knowledge infrastructure would have to provide the superset of such functionalities across different paradigms.

3.3. Object-oriented model of scientific knowledge

Developers of knowledge infrastructures need to find an adequate structure for representing scientific knowledge within their information system. As the previous sections have shown, this is a challenging problem: neither scientists nor philosophers of science have reached consensus about the characteristics of knowledge and no current interpretation is without fundamental problems. As a consequence, it is far from obvious what the elements of scientific knowledge are and what their properties and relationships are. Moreover, choices about the representation of knowledge may limit the types of scientific paradigms which a knowledge infrastructure can support.

Unfortunately, the representation schemes for scientific knowledge used in traditional journals and most current online infrastructures have significant epistemological shortcomings. Paper journals, for example, adopt a very simple structure which does not distinguish different types or characteristics of scientific knowledge. As a consequence, it is not possible to meet basic information needs of the scientists – such as providing information about the set of current theories in a field.

The goal of this chapter is to identify a structure of scientific knowledge which is epistemologically more adequate in an online environment. In information systems terminology, formal and semi-formal representations of the structure of a domain which are independent of the choice of a specific information technology are called conceptual models. In this section, therefore, a conceptual model will be developed which describes the structure of scientific knowledge. The model will focus on components of knowledge, their relationships and attributes which need to be represented in a scientific knowledge infrastructure. In addition, the analysis will also include dynamic aspects and describe functionality which is needed to support knowledge infrastructures.

In the following, the structure and applications of a conceptual model will be analyzed first. Then a meta model and finally the conceptual model of scientific knowledge will be developed. The purpose of this section, therefore, is to develop an information model for knowledge infrastructures which is based on the theory of knowledge and which can be used as the basis for the development of a scientific knowledge infrastructure. Such a model has the potential of significantly reducing the trial and error process on which current knowledge infrastructures are embarking.

Thus the goal of the conceptual model to be developed in the following is to articulate the components and relationships of scientific knowledge in a way that corresponds to the implicit and shared understanding of researchers about scientific knowledge. The following sections aim to identify a maximum common denominator among the many theoretical and practical-implicit conceptualizations of scientific knowledge.

3.3.1. Approach

In the field of information systems, a variety of methods have been developed for the development of conceptual model which describe the structure of a target domain (Hars 1994, pp 28-34). Examples are the Architecture for Integrated Information Systems (ARIS) (Scheer 1998a, 1998b), object-oriented approaches such as OMT (Rumbaugh et al. 1991) or UML (Booch, Jacobson & Rumbaugh 1998), and information engineering (Martin 1990). At the core of all these methods are models - systematic, abstract descriptions of relevant aspects of a domain.

Every model involves a ternary relationship which relates two systems – 1) an object system and 2) a model system – to one another for 3) a specific purpose. When object system and purpose are obvious, the term 'model' is often used ambiguously instead of the more precise 'model system' (see also Grochla 1974 p. 21; Bernzen 1990, p 428). For a more detailed discussion of the term 'model' see Hars (1994, pp 7-12). A classification of different types of models can be found in Bunge (1998a, p 478).

Modeling methodologies typically use several different types of models. The predominant model types for describing the conceptual structure of a domain are data and object models. Such models describe the concepts which are relevant in a domain as well as their attributes and relationships. Object models add a description of the behaviors associated with instances of the concepts. Besides these types of models, other types exist which describe the dynamic behavior of elements in a domain (e.g. process models), organizational structures (organization models), etc.

The subsequent analysis will focus on the conceptual structure. An object-oriented approach is preferable over a data-oriented approach, because object models are extensions of data models and because of their larger expressive capability. Many different object-oriented methodologies exist. Among them, the Unified Modeling Language (UML), which integrates three leading object-oriented modeling methodologies, is rapidly becoming the standard for object-oriented analysis and will therefore be used subsequently. The analysis will focus on the most central of UML's many model types: the class diagram.

Several approaches exist for the development of class diagrams. Most methodologies begin with an analysis of texts or statements describing the domain. Methodologies for conceptual model development using text analysis are described in: Nijssen & Halpin (1989), Kristen (1994) and McDavid (1996). From the texts, candidate class, attributes and relationships are extracted. Tentative models are then built, analyzed and refined. When different statements about the same domain are available, tentative models are developed in parallel for each perspective. Subsequently the models are integrated. This requires the identification of differences and the removal of inconsistencies (Hars 1998; Wagner 1989). Subsequently this approach will be adopted for the development of the conceptual model of scientific knowledge. Class diagrams will be built by analyzing the characterizations of scientific knowledge provided by eminent philosophers of science.

Before a conceptual model of scientific knowledge can be developed, additional criteria for its evaluation need to be determined. Criteria for adequate representations have been defined in various fields such as database management, where some researchers have begun to formalize requirements into Generally Accepted Modeling Principles in analogy to Generally Accepted Accounting Principles (Rosemann 1996). Another field is artificial intelligence where criteria for the evaluation of formal representations have been developed. The following criteria need to be distinguished:

- **Clarity** (Rosemann 1996)**:** Definitions should be maximally clear. Examples and counterexamples should be given. Underlying assumptions need to be made explicit.
- **Consistency:** A model needs to be internally consistent. Circularity in definitions must be avoided. For external consistency, terms should conform to common usage. In addition, the representation should be structurally and behaviorally adequate (Rosemann 1996).
- **Coherence:** Relationships between categories need to be carefully defined. Overlap between different categories needs to be minimized and synonyms need to be eliminated (e.g. via careful naming or lexical analysis algorithms (Hars 1998).
- **Extensibility:** It must be simple to add new elements to the model.
- **Reusability:** A model should be designed to be reusable in different contexts.
- **Metaphysical adequacy:** "A representation is called *metaphysically adequate* if the world could have that form without contradicting the aspect of reality that interests us." (McCarthy & Hayes 1969, p 27) The representation of reality as consisting of collections of particles which interact through forces between particles, for example is often metaphysically adequate. "Metaphysically adequate representations are mainly useful for constructing general theories." (McCarthy & Hayes 1969, p 27)
- **Epistemological adequacy:** Metaphysically adequate representations are not necessarily adequate for representing those aspects of reality that are important for an information processing system. It is difficult for example, to represent common-sense knowledge such as "dogs chase cats" or "John is at home" as collections of particles linked via forces. Therefore "a representation is called *epistemologically adequate* for a person or machine if it can be used practically to express the facts that one actually has about the aspect of the world." (McCarthy & Hayes 1969, p 27)
- **Heuristic adequacy** (McCarthy & Hayes 1969, p 27)**:** McCarthy & Hayes introduce this variant of adequacy to distinguish cases in which a representation is epistemologically adequate but not suitable for supporting reasoning and inference processes.

- **Parsimony:** A representation needs to avoid unnecessary overhead. It should distinguish between original, axiomatic knowledge and knowledge which can be derived through mathematical or logical operations from already existing knowledge.

Before a detailed conceptual model of scientific knowledge can be developed an appropriate level of abstraction needs to be identified. Conceptual models are often based on a highly generic meta model from which specific models can then be derived (Hars 1994, pp 11-12). In the following, therefore, first a meta-model of scientific knowledge will be developed. Then various epistemologies will be analyzed, integrated and extended.

3.3.2. Meta model

A model of scientific knowledge can be developed on several levels of abstraction. The most abstract model would represent all kinds of knowledge as instances of a single, generic class. This class will be called a 'unit of knowledge'. On this abstract level, knowledge can be represented as a network of units of knowledge which are linked through some relationships (see Figure 13a). The term unit of knowledge includes concepts which are at the core of traditional knowledge infrastructures such as 'article' or 'document'. It also covers other forms of scientific knowledge such as texts, equations, formulae, graphs etc. Furthermore, a unit of knowledge may refer to epistemological units such as a theory, a problem description, etc.

Refinement of this abstract class diagram occurs by distinguishing different types of nodes and/or different types of relationships. Most knowledge has some direct relationship to people. For example, persons create knowledge. Therefore a class called 'person' and a relationship class such 'authored-by' is introduced. In addition, it must be possible to comment and evaluate every node. Therefore, 'comments' are introduced as an additional subtype of 'node'. Both information about persons and comments themselves are units of knowledge. The arrow indicates this generalization relationship. These refinements are shown in Figure 13b.

The model can be further refined by differentiating different types of relationships between contributions such as 'aggregated-to', 'caused-by' etc. This requires an additional class called 'relationship type' (see Figure 13c). Different relationships between contributions can then be represented as a ternary relationship between one contribution which is at the source of the relationship, one contribution which is at its destination and the specific relationship type. In object-oriented notation, relationships which are not binary require the introduction of a relationship symbol (diamond) and an associated relationship class ('related to'). In addition, different types of units of knowledge may need to be distinguished. This leads to the introduction of a class called 'unit type'. The diamond at the end of the relationship indicates that units of knowledge can be aggregated into 'unit type' categories.

This top-level model can subsequently be refined by identifying subcategories of nodes and defining additional relationship types.

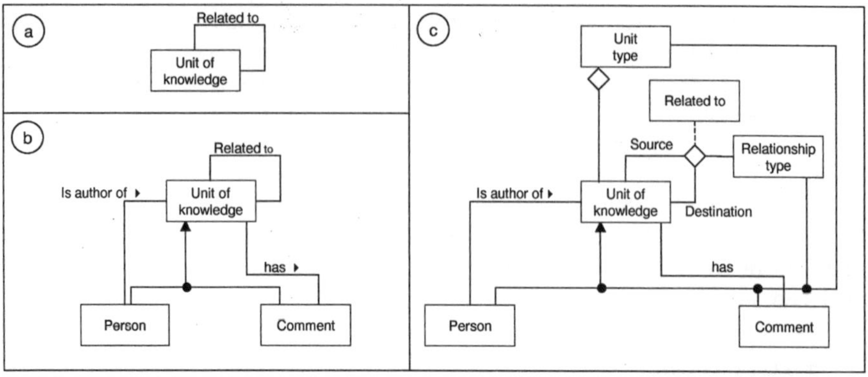

Figure 13. Conceptual models of scientific knowledge on different levels of abstraction

3.3.2.1. Generic attributes

Besides the network of relationships in which a unit of knowledge can be embedded it is necessary to identify its attributes. Librarians have long grappled with the problem of how to meaningfully describe books and a variety of other types of knowledge. Library systems, therefore, can provide the basis for the identification of properties. In recent years, an international library initiative has developed a base set of meta data for library resources, the 'Dublin core'. It consists of 15 "base elements for descriptive metadata" (Weibel & Hakala 1998; N.N. 1997) (see Table 15).

Table 15. Dublin core meta-data elements

Content	Intellectual property	Instantiation
Title	Creator	Date
Subject	Publisher	Type
Description	Contributor	Format
Source	Rights	Identifier
Language		
Relation		
Coverage		

Source: Weibel & Hakala 1998.

Some of these elements capture information which describes a relationship. Examples are 'creator' and 'publisher'. Both are not properties of a unit of knowledge in a narrow sense; their existence is quite independent from the contribution. Such attributes are called 'derived attributes'. In object orientated notation, a forward slash can be used to indicate such cases. A list of authors, for example, can be derived from the 'authored by' relationship which exists for a knowledge unit. Some properties may contain multiple values while others (e.g. identifier) consist of a single value. Several elements need special consideration: the language element is only necessary for the subset of linguistic knowledge and does not apply to e.g. formulas, pictures, etc. The relation element is a placeholder for different kinds of relations; it can be eliminated from this model because relationships are defined explicitly. The coverage attribute describes the spatial or temporal context of knowledge; its relevance has been hotly debated. As a consequence, most attributes of the Dublin Core are generic attributes of a knowledge contribution (see appropriately called 'presentation'. This leads to five core activities (create/edit, integrate, evaluate, search, present,) which will be further examined in the following. Two additional categories of activities are relevant: the generation of knowledge and its application (see Table 17).

Table 16. Generic properties and activities for a unit of knowledge

Unit of Knowledge			
Properties			Activities
Title	Source	Identifier	Edit
Description	Relation	Date created	Evaluate
Content		Date modified	Search
	Creator	**Type**	Present
Subject	Publisher	**Format**	Transform
Coverage	Contributor	Version	Integrate
Language	Rights	History	Summarize
Quality			Generate
			Apply

The Dublin Core has been developed from an information retrieval perspective and therefore does not describe all attributes of a unit of knowledge. An additional 'content' attribute is required which contains e.g. the full text of an article. Not all derived attributes have been listed. Additional attributes may be required for performing certain activities on a contribution.

3.3.2.2. Knowledge-related activities

A second aspect of analysis arc the activities which can be performed on units of knowledge. Information tcchnology may change the way in which these activities are executed. For example, repetitive and well-structured activities such as indexing, transformation of formats, etc. may be automated. Another avenue for the use of information technology is the interactive manipulation of knowledge where the computer processes knowledge according to the instructions of a researcher. Already in the first half of the 20th Century, Bush argued that:

> "Whenever logical processes of thought are employed -- that is, whenever thought for a time runs along an accepted groove -- there is an opportunity for the machine." ... "We may some day click off arguments on a machine with the same assurance that we now enter sales on a cash register." (Bush 1945)

Information technology may also be useful for coordinating distributed processes through work-flow automation. Such approaches are already in use for reviewing. Finally, information technology allows one to reuse knowledge by providing access to best practices, templates and software wizards.

In the following, the main activities related to a unit of knowledge will be identified. In object-oriented analysis, activities are represented as methods of classes. Instances of classes interact with one another (and with the user) through such methods. Subclasses inherit the methods but may introduce additional methods or may change the inherited methods. In the latter case, the subclasses 'override' the inherited method with their own implementation of the method. In a section further below, therefore, the analysis will be extended to the subclasses of 'unit of knowledge' such as 'concept', 'theory', 'method' etc.

The traditional publishing process may be used as starting point for the identification of activities which need to be provided for a unit of knowledge. The reengineered process (See Figure 5 on page 32 above.) begins with the *creation* and *editing* of a unit of knowledge. This is followed by *review* and *reading*. Several activities from the traditional process are no longer relevant: copy-editing may be subsumed under editing. Printing can be disregarded as on-screen viewing, annotation and editing become more prevalent. Some of the activities need to be broken down further. For example, creation/editing implicitly involves the inclusion of references to other articles. In an online environment the *integration* of new knowledge into the existing body of scientific knowledge may increase in importance and therefore will be treated as an additional activity. Similarly, an important additional aspect of reading is *searching* for relevant units of knowledge. The term 'reading' is user-centric. The service which a knowledge unit renders to the reader is more appropriately called 'presentation'. This leads to five core activities (create/edit, integrate, evaluate, search, present,) which will be further examined in the following. Two additional categories of activities are relevant: the generation of knowledge and its application (see Table 17).

Note that such an analysis does not require the introduction of computer technology. Even in the traditional publishing system, formal guidelines and procedures exist for most of these services. In the current publishing system, for example, much literature describes how to write scientific articles. Explicit standards have been developed for review and for hundreds of years libraries have developed classification systems to facilitate search. One of the tasks during the implementation of information technology is to translate these existing formalizations into code. Additional activities that need to be considered are the generation and the application of knowledge:

3.3.2.2.1. Create & edit

Current traditional and online knowledge infrastructures provide little support for creating and editing knowledge. Authors write manuscripts locally using word processing software, graphics programs, formula editors. Although manuscripts for electronic journals are usually submitted in electronic form they require additional processing before they are placed online. This may include some editing and conversion into the appropriate electronic format such as HTML or PDF. In the near future it may also include XML markups. Only few infrastructures provide direct upload capability. An example is the Stanford Encyclopedia of Philosophy (see section 2.2.3.2.).

Table 17. Activities related to knowledge

Activity	Issues
Create/edit	Create/modify/delete knowledge in different formats and notations Markup & structuring of knowledge Predefined templates for specific types of knowledge Procedures & guidelines for evaluation Collaborative editing Version control ⇨ Integrate/link with other knowledge ⇨ Evaluate knowledge ⇨ Present knowledge
Integrate	Identification of dependencies Automatic categorization Removal of inconsistencies Linking into common framework Procedures and guidelines for manual integration Alert users of significant changes
Evaluate	Workflow support for review process Track reviewer performance Summarize knowledge Collaborative filtering Calculation of quality measures Procedures & guidelines for evaluation Network analysis Alerting users of significant changes
Present	Dynamic presentation depending on user preferences Anticipate user preferences Increase/decrease level of abstraction Custom assembly of knowledge ⇨ Summarize knowledge
Search	Query formulation Query execution, including optimization Interactive (assisted) browsing Generate and manage trails Monitor user, generate user profile
Generate	Generate variants Paraphrase knowledge
Apply	Analyze context Populate model Validate results Apply knowledge

The focus on dissemination and the lack of support for editing has significant negative consequences. It inhibits the integration of new knowledge into the existing body of knowledge. It increases publishing delays and therefore reduces the suitability of an infrastructure for collaboration.

Knowledge infrastructures can provide functionality to support editing and creation of knowledge in several ways: on the most basic level, mechanisms are needed to create knowledge in a specific notation. This may include basic text editing capabilities, formula editors, general diagramming tools etc. In these areas, knowledge infrastructures can hardly compete with existing products. Infrastructure-specific editors, however, can provide specific markup functionality which enables an author to identify and classify key elements of his contribution. The markup can be used subsequently for search, integration and evaluation. In addition, custom editors need to provide services for the more structured types of scientific knowledge. For example, a graphical editor for structural equation models which could be interfaced with statistical software packages would be useful for empirical research. Another feature which custom editors may provide are templates for specific types of knowledge. Different templates might exist, for example, for different types of quantitative and qualitative research approaches. These templates standardize descriptions, ensure interoperability between different contributions and ensure that contributions are complete. Another advantage of templates is that they may reflect best scientific practices. Furthermore editors may coach authors during knowledge creation. They may contain procedures and guidelines for typical scientific issues. These may be provided as context-sensitive manuals or through wizard-like help applications. In contrast to current text processors, editors are needed which provide support for the thinking process rather than just supporting the entering and changing of thoughts in an electronic medium. An editor, for example, may generate questions for the type of issue which is being addressed. It may list exemplary tables of content to help the author in finding the right structure for his contribution.

More advanced editing systems may include support for group editing and version control. In addition, custom editors need to provide links to additional functionality for integrating, evaluating and presenting knowledge.

3.3.2.2.2. Integrate

A key goal of science is the accumulation of a growing body of scientific knowledge. Although scientific progress is not always cumulative (Kuhn 1996), scientists spend much effort to show how their work relates to previous research. In the paper-based environment only few mechanisms exist for integrating units of knowledge. References serve as (one-directional) links between units of knowledge. They allow to trace the origin of scientific ideas. The physical co-location of units of knowledge in reference books and special issues of journals are other examples of rudimentary integration. As was shown in section 3.1.2. on page 84, the nature of paper significantly reduces the potential for integration. As a consequence, traditional science only has a small number of procedures and algorithms for systematically integrating knowledge.

Information technology significantly increases the potential for integration. Integration requires the identification of dependencies, the removal of inconsistencies between different findings and the linking of knowledge into a common framework. A variety of tools can be utilized for integration. For natural language text, an analysis of synonyms can be utilized. For concepts, generalization hierarchies and taxonomies can be used. Integration may include the triggering of actions within the body of knowledge once certain relationships have been established. Another aspect of integration services is classification. Establishing categories of similar phenomena and identifying the associated categories for a phenomenon are additional aspects of integration.

3.3.2.2.3. Evaluate

An important characteristic of the scientific body of knowledge is its quality. Many different approaches exist for evaluating the quality of scientific knowledge. Currently the predominant mechanism is peer review, which is performed manually. It heavily relies on the reviewers' expertise and on their preference structures.

When moving towards an online environment manual procedures can be streamlined. A workflow system may be used to track review progress and to handle the communication between the parties involved. The review process may be refined to systematically track revision changes and ensure that authors attend to each issue raised by a reviewer. This intensifies the interaction between author and reviewers. Additional mechanisms may need to be added to resolve conflicts and to avoid stalemates when opinions diverge. Computer support, however, provides the additional capability of tracking reviewer performance, of providing templates for review which may increase the consistency of reviews. Such templates may be fine-tuned for different types of methodologies. In addition, IT provides an opportunity for developing automated and semi-automated procedures for evaluating scientific contributions. Algorithms may analyze the network of relationships into which the unit of knowledge is embedded. Search algorithms may provide the reviewer with prior contributions that have not been cited or included in the research. Another way in which information technology can assist the evaluation process is by retrieving reviews for similar contributions which the reviewers may use as a guideline. The result of evaluation leads to information about the quality of a contribution which then can be regarded as an attribute of a unit of knowledge. The structure of the attribute, however, depends on the evaluation approach. It may range from a binary 'accepted'/'not accepted' attribute to a multidimensional attribute which reflects the opinions of many reviewers.

Another aspect of evaluation is summarizing knowledge. This activity is already present in current scientific articles which contain three levels of summarization. The body of an article represents the basic level of summarization (no summarization). The next level is formed by the abstract which usually accompanies an article and the highest level of summarization is the title. Depending on the type of knowledge, different mechanisms exist for summarizing. Elaborate algorithms exist, for example, for maps and much research currently focuses on the creation of automated summaries of natural language texts, images and videos. Thus different types of units of knowledge may be associated with different mechanisms for the creation of summaries.

3.3.2.2.4. Present

In the paper environment the author entirely determines the presentation of knowledge at the time of writing. In an IT-based environment the presentation need not be fixed but can be generated at the time of reading. For example, an author may store data explicitly with a text instead of showing a few diagrams. The reader may then decide on which level of abstraction and in which format the data should be shown and whether it should be presented as tables or graphs. A key technology which supports alternative presentations of knowledge is called Extended Markup Language (XML). It is currently being embedded in major internet browsers. Another low-level example of a presentation service is Java's layout manager.

Variety in presentation need not be limited to data. Linear text is a compromise which balances the information needs of the expert with those of the student. In online text, information about the addressees or the level of detail can be added to a section. This information can then be used to assemble a text dynamically to the needs of each reader.

Thus, presentation services need to be able to dynamically change the layout of text and graphics depending on reader profiles and preferences. This includes hiding and displaying sections of texts, aggregating texts and data, and selecting appropriate formats for presentation. Presentation may include functions which generate an outline of a text or argument which a reader can use for navigation. It may support trails which help the reader acquire knowledge. Presentation services require changes during editing: authors need to specify types of presentations, mark up argument structures and provide approaches for navigating the knowledge. Thus, a unit of knowledge can be associated with a variety of different forms of presentations. In addition, rules and standards may be needed to define how contributions should be presented in certain contexts.

Presentation services provide many challenges and it may take some time until the linear and static hypertext models of presentation are replaced by dynamic presentation services. The separation between the description of knowledge and its presentation, however, is an area with many opportunities for knowledge infrastructures.

3.3.2.2.5. Search

Search is an important activity and a standard service in knowledge management infrastructures. A search service takes search criteria and identifies candidates among the units of knowledge which match these criteria. A search may include algorithms which estimate the relevance of an article by analyzing its citation pattern and its other characteristics. An analysis of citation and readership patterns may allow to spot 'classic' papers early on (Price 1986, p.108). For challenges of bibliometric analysis of hyperlinked documents on the web, see Harter & Ford (2000).

3.3.2.2.6. Generate

After a new theory has been introduced, much research will follow to extend and refine the scope of the theory. Scientists will attempt to add additional factors to the model or introduce additional complexity by refining variables, adding uncertainty etc. In short, many ways exist in which a theory can be extended. The purpose of generation is to systematically create extensions of a knowledge unit by performing standard operations on this knowledge. In current research practice, generation is often used as the basis for a new contribution: a researcher adds additional factors to a known problem and shows how the resulting problems can be solved. This is what Kuhn (1996) calls 'normal science'. However, systematic generation of variants is not significant within the current practice of management science. Similar to the other activities generation may be based on software or may be performed manually.

3.3.2.2.7. Apply

An important aspect of knowledge is its application. As long as explicit scientific knowledge has been limited to paper, direct application was difficult. Due to the fragmentation of the scientific body of knowledge, the direct application of the insights contained in articles was rarely possible. The application of knowledge required either detailed instructions in the form of text books etc. or direct support from a scientist who was familiar with the topic. With the advent of IT, this situation is changing. As knowledge becomes dynamic, units of knowledge may be applied automatically. Private investors may use the software version of a novel investment theory to make purchases on the stock exchange. A manufacturing company may download a recently developed scheduling algorithm and use it to generate their production schedule. Thus services are needed which are capable of executing dynamic knowledge.

3.3.3. Analysis of epistemologies

Knowledge infrastructures need to refine the generic structure of knowledge described above. They need to be able to differentiate different categories of scientific knowledge. These categories need to be specializations of the generic 'unit of knowledge' category. In the following, established theories of knowledge will be analyzed to identify these subcategories. The analysis will be complemented by analyzing texts on research methodology which adopt a practical perspective.

In the philosophy of science, many authors have built comprehensive systems of scientific knowledge and characterized the different components of knowledge, their attributes and relationships. In the following, the systems of four different authors will be analyzed to determine their implied structure of scientific knowledge. First, Popper's Logic of Scientific Discovery will be analyzed because of its high influence on the practice of science in the second half of the 20th century. Subsequently Nagel's 'Structure of Science' will be evaluated. It is a systematic analysis of the main components of scientific knowledge. Next, Dubin's 'Theory building' will be discussed. Among the four epistemologies it is the most practically oriented. The fourth conceptual system to be analyzed is Bunge's 'Philosophy of science'. It is the most recent, detailed and extensive of the four epistemologies.

In each case, the main components of knowledge distinguished by the author will be identified. After determining their relationships, a class diagram will be created using object-oriented notation. However, the resulting class diagrams can only be an approximation. Unfortunately none of the four epistemologies have been completely formalized. Key concepts are not always precisely defined; small inconsistencies exist between different parts of the texts; some relationships are not explicitly defined but must be logically deduced by the reader. Therefore, the resulting class diagrams cannot be fully objective; to a degree, they depend on the author's interpretation.

3.3.3.1. Popper: Logic of scientific discovery

Popper interprets the theory of knowledge as the "analysis of the method or procedure peculiar to empirical science" and therefore equates epistemology with a "theory of the empirical method." (Popper 1968, p 39) As a consequence, the structure of knowledge is not his primary concern – rather the context of scientific knowledge creation, refinement and refutation. He does not find the difference between scientific and other knowledge in its structure, but rather in the method and process which it is embedded in. He is convinced that scientific knowledge does not rest on solid foundations (Popper 1986, p 111).

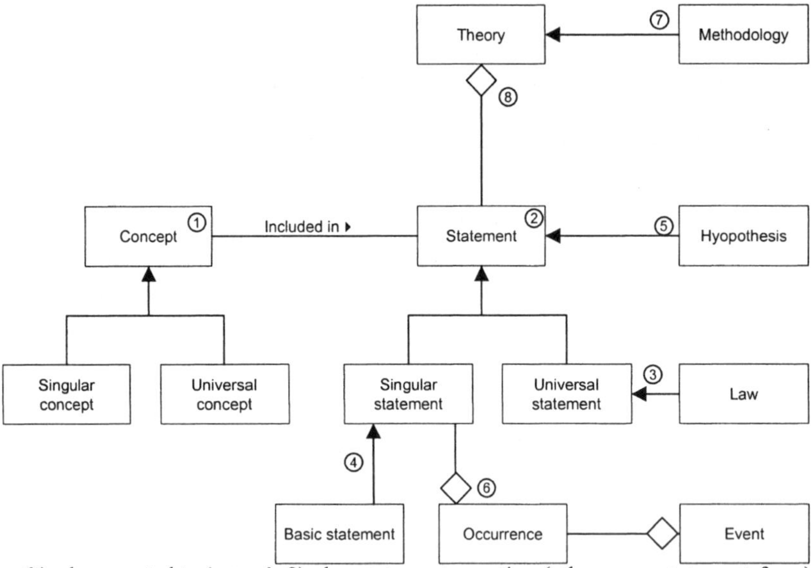

1) p 64, also equated to 'name'; 2) also statement-equation (others use: statement-form), p 17, p 73; 3) p 59, p 63 but also: law is a special kind of theory; p 68f; 4) subsets are 'basic statements' which may falsify a theory (p 106, p 43), other subset: initial conditions (p 59); 5) Hypothesis = empirical statement, p 27., p 72; 6) occurrence and event are elements of the 'real' world; 7) p 49; 8) pp 71-75. Popper occasionally equates theories with statements (e.g. p 59)

Figure 14. Popper's meta structure

As with most other conceptualizations of scientific knowledge, universal and singular statements are at the core of his conceptual structure (see Figure 14). This includes statements with variables which he calls statement-equations.

Statements make reference to concepts which may either be singular or universal. An occurrence reflects the meaning of a singular statement. It is the real phenomenon on which the statement is based. While this phenomenon may be described in different ways, all such statements describe the same 'occurrence'. Singular occurences which are similar, except on some dimensions such as time and space, can be aggregated into 'events'. In this way Popper builds the bridge between idea and reality. Popper does not clearly describe the relationship between 'occurrences' and concepts. However, he strongly opposes the objectivization of experience and observation. Both cannot be regarded as truthful but must submit to testing as all other elements of scientific knowledge.

3.3.3.2. Nagel: Structure of science

Nagel (1961) sets out to examine the logical structure of scientific explanations. He identifies four patterns of explanation: deductive, probabilistic, functional/teleological and genetic (the term 'genetic' explanations is used here in the sense of 'historical' explanations). He then analyzes the characteristics of scientific statements. He begins with laws and empirical generalizations and proceeds to examine the 'cognitive status of theories'. Experimental laws consist of a single statement. They are empirically determinate. All non-logical concepts which occur in the law must be observable.

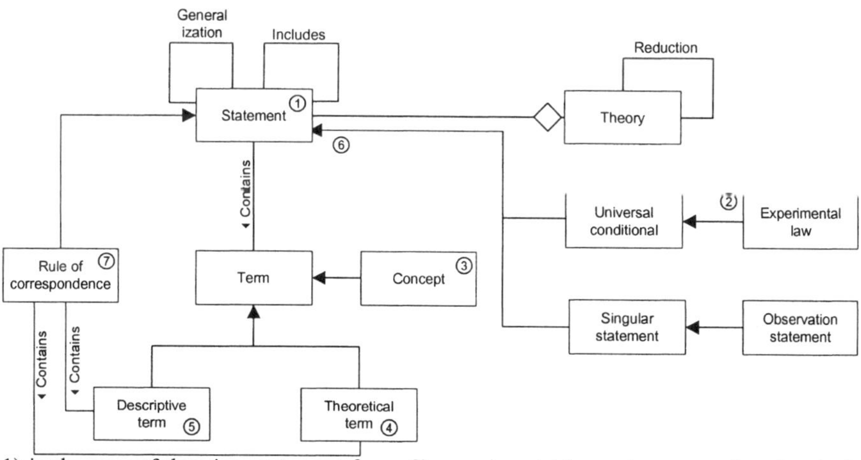

1) in the case of theories: statement-form; 2) experimental laws always consist of a single statement; 3) role of concepts not clearly described 4) constraint not shown: theories must contain theoretical terms; 5) constraint not shown: experimental laws contain descriptive terms but no theoretical terms; 6) partial list of specializations; 7) synoym: operational definition, pp 93, 97-105.

Figure 15. Nagel's meta structure

In addition, they may be derived by induction from empirical generalizations (Nagel 1961, pp 83-90). These three characteristics set laws apart from theories which are systems of statements. They consist of three elements: an abstract calculus, a mapping mechanism and an interpretation (Nagel 1961, pp 90-97). He also provides a theoretical basis for the absorption of one theory by another, which he calls 'reduction'. Subsequently, Nagel examines key problems in different areas of the sciences. Beginning with mechanics, he analyzes the epistemological problems associated with Newton's Laws addressing such issues as whether his law of motion has empirical content or merely has the status of a definition. Finally, he analyzes methodological and epistemological problems of the social sciences and of historical inquiry.

Nagel provides compelling insights about the nature of theories and scientific knowledge. He consistently uses a small number of core concepts (e.g. statements and terms) for his argument. The inner structure of statements including the nature of scientific concepts are not analyzed in detail. The same applies to other elements of scientific thought such as hypotheses and problem statements. A major advantage of Nagel's analysis, however, is that it extends beyond the traditional reference sciences (physics, mechanics) to include the social sciences and history.

3.3.3.3. Dubin: Theory building

Motivated by the need "for more specific operational linkages between theory and research, that could be put to everyday use by a practicing researcher" (Dubin 1978, p xi), Dubin has developed a detailed conceptualization of scientific theories which differs significantly from Popper's and Nagel's conceptualizations. It consists of seven components (see Table 24 on page 157).

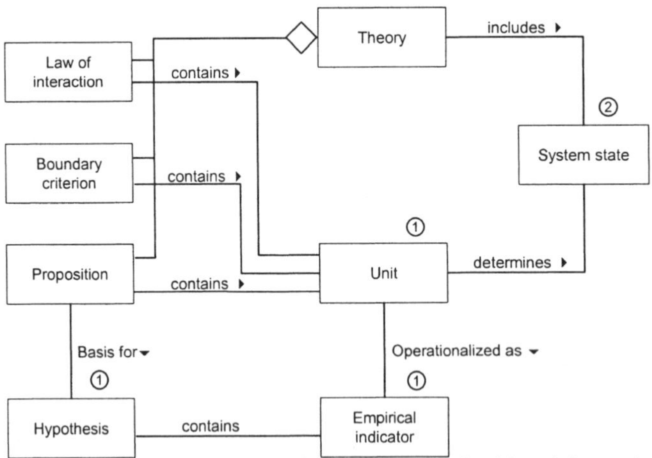

1) The relationship between unit (empirical indicator, hypothesis) and theory is implicit; 2) System states might require state description statements (not mentioned by Dubin).

Figure 16. Dubin's meta structure

The main components are 'units'. Units are 'properties of things'. They can be observational or theoretical terms and are equivalent to the term 'concept' used by Popper and Nagel. Besides units, Dubin distinguishes several different types of statements. Laws of interaction are statements which reflect dependencies between units. Another type of statement describes internal and external boundaries of a theory. Dubin recognizes that boundary statements and interaction statements may overlap (1978, p. 132). He observes that the behavior of a system may vary depending on its overall state. An example is a market system which exhibits the states of perfect competition, imperfect competition or monopoly. Therefore, Dubin requires the identification of those persistent system states in which the laws of interaction are valid. Dubin's system states cannot be partial. Moreover, theories do not require the existence of states. Propositions and hypotheses are special kinds of statements which relate the units of the theory. Propositions follow from the laws of interaction and hypotheses are operationalizations of propositions. Empirical indicators are reflected in the meta structure as measures which are special types of concepts.

3.3.3.4. Bunge: Philosophy of science

In his 'Philosophy of Science', Bunge (1998a, 1998b) presents a detailed analysis of scientific knowledge. His conceptualization of knowledge (see Figure 17) is based on the distinction between ideas and facts where ideas are then broken down into problems, factual hypotheses, laws and theories. Factual hypotheses are the basic building blocks of scientific ideas. The label 'factual' is misleading because it is used as an opposite to observational hypotheses which are empirical statements. Factual hypotheses represent corrigible statements concerning "facts that are as yet unexperienced or in principle unexperienceable." (Bunge 1998a, p 254) Factual hypotheses often cannot be extracted from data, but require a creative leap. When factual hypotheses are confirmed, they may advance to the level of a scientific law (Bunge 1998a, p 355). Hypotheses – both factual and observational – are assembled into theories (Bunge 1998a, pp 253-259). Bunge stresses the importance of problems which he describes as the "springs of science" (Bunge 1998a, p 187) and outlines a calculus of problems. Data is generated by scientific experiences which include observation, measurement and experiments. Data is frequently linked to hypotheses in the context of a theory.

Bunge clearly distinguishes between logical and epistemological aspects of the elements of knowledge. In his model (Figure 17) this is reflected in the distinction between 'hypothesis' as a special kind of 'formula' (logical aspect) and 'hypothesis' as a special kind of 'scientific idea' (epistemological aspect). It is also reflected in his terminological choices: he prefers the term 'formula' over the logical terms 'statement' or 'statement-form'. He emphasizes that scientific knowledge is rarely completely formalized: "complete formalization [of a theory] is rarely attempted outside the formal sciences [i.e. logic, mathematics]." (Bunge 1998b, p 548). A similar position is adopted by Popper (1968, p 379) who strongly argues against the formalization of science through "pre-established language systems." This would be "the end of science". Therefore a representation of scientific knowledge must be able to incorporate imperfect scientific knowledge. As a consequence, it is not possible to require that scientific knowledge be formulated in a variant of first-order logic. While knowledge infrastructures should support logic, they need to support other approaches for the representation of knowledge including the widely used form of natural language text.

Bunge provides many criteria for the evaluation of his categories of knowledge and identifies many misconceptions. The categories shown in Figure 17 form the top level of Bunge's epistemology. All concepts are broken down into much further detail in his work.

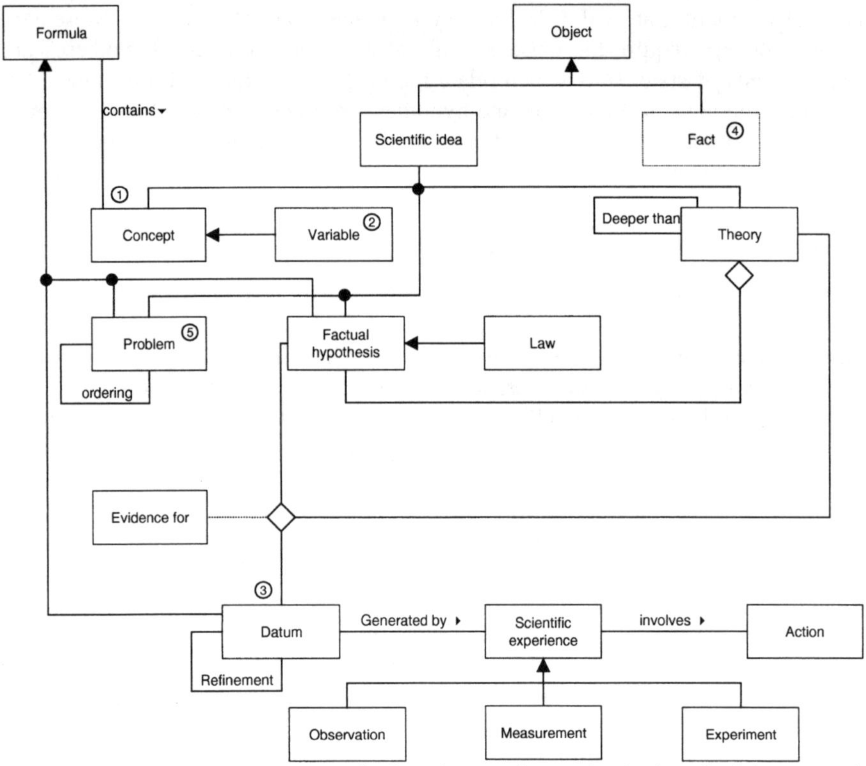

1) Relationship between concepts and other elements is not clearly described (except variables: p 351) 3) At times Data are described as observational hypotheses; 4) facts are real but never directly accessible (instead data are needed which are interpretations of facts) therefore they are not part of the meta strcuture; 5) separate calculus is proposed for problems. 6) variable: p 347-355.

Figure 17. Bunge's meta structure

3.3.3.5. Babbie, Emory: Epistemologies in scientific practice

The previous four conceptualizations of scientific knowledge have been based on theoretical arguments. It is not certain that these analyses accurately reflect the elements of scientific knowledge which are of main concern in scientific practice. An alternative approach for identifying the elements of scientific knowledge is the analysis of texts which deal with the practical aspects of conducting science. Texts on research methodology can be used as the basis of such an analysis because they describe the issues which arise during research in detail. They are typically structured around different types of scientific knowledge.

In the field of information systems, several methods for the analysis and development of meta structures have been developed Hars (1994). Their approaches are similar: a thorough analysis of the terminology in a specific domain on the basis of documents or interviews usually forms the basis (McDavid 1996). Natural language statements can systematically be analyzed to identify the predominant types of concepts and their relationships. Kristen (1994, pp 103-106) describes a detailed approach which starts with the creation of a list of structured sentences. From these sentences, candidate objects are identified. Synonyms are identified and removed. Homonyms are replaced by distinct terms. In a final step, relationships between objects are identified. This is combined with a representation scheme, which often is a variant or extension of the Entity Relationship Model (Chen 1976).

Table 18. Terms related to research methodology

Babbie (1994)		Emory (1984)
Reality ×	Study	Study
Objectivity, subjectivity	Conceptualization	Question
Observation	Research proposal	Concepts✓
Fact✓	Dimension	Definition
Law	Indicator	Research design
Theory✓	Definition	Method
Concept✓	Research purpose	Hypothesis✓
Variable✓	Illustration	Proposal
Proposition✓	Questions	Research plan
Hypothesis✓	Statement	Measure
Research design	Questionnaire	Scale
Scale	Data✓	Experiment
Typology	Terminology	Quasiexperiment
Experiment	Sampling frame	Data✓
Survey	Cluster	Survey
Field study	Variable✓	Observation
Paradigm	Problem✓	Survey instrument
Model	Simulation	Response
Reason	Code	Sample
Criterion	Issues	Test
Error	Examples	Research report
Evidence		
Conclusion		
Measurement		

Following this approach, the terminology used in two texts on research methodology (Babbie 1994; Emory 1984) were evaluated. To identify the potential elements of the meta structure, the tables of contents of these books were analyzed. In addition Babbie (1994, p 48-49) explicitly lists 10 terms relevant for theory creation. All candidate terms are shown in table Table 18.

Some of these terms, such as 'reality', 'objectivity and subjectivity' etc. do not represent elements of knowledge but rather specific scientific issues. These terms are marked with a ×. The remaining concepts have significant overlap. For example 'observations' and 'facts' are closely related. The same applies to 'evidence' and 'data'. Table 19 classifies similar terms into groups. The main term of each group is used as heading.

The groups of concepts identified in Table 19 are closely linked. Formal statements, for example are used to express facts, to define concepts and to describe potential problems. Hypotheses, axioms and propositions are special kinds of statements. A theory consists of sets of statements. Facts are aggregated into data. Concepts are often aggregated into taxonomies, typologies or frameworks. We use the term framework instead of typology or taxonomy because it has the widest scope of the three terms. Variables and measures are special types of concepts. Theories may be used to explain certain statements.

Table 19. Clustering of terms

Concept	**Theory**	**Measure**	**Other**
Variable	Law	Scale	Definition
Conceptualization	Paradigm	Survey instrument	Measurement
	Model	Dimension	Illustration
Statement		Indicator	Criterion
Conclusion	**Typology**	Code	Sampling frame
Reason	Terminology		Cluster
Proposition	Taxonomy	**Study**	
Hypothesis		Experiment	
Assumption	**Problem**	Quasiexperiment	
	Issues	Simulation	
Fact	Question	Survey	
Observation	Research purpose	Field study	
Evidence	Proposal	Research report	
Response	Research proposal		
Examples	Error		
Data			
Sample	**Questionnaire**		
	Test		**Not relevant**
Method			Reality ×
Research design			Objectivity ×
Research plan			Subjectivity ×

3.3.3.6. Synthesis

The epistemologies discussed above can be used as starting point for developing a comprehensive model of scientific knowledge. This requires the identification of their commonalties and the resolution of their differences. In the following, a model of scientific knowledge will be assembled from the four epistemologies and the two publications on research methodology. The resulting model is shown in Figure 18. To synthesize the models, first, those elements of knowledge which are at the core of the epistemologies will be identified. Next, their relationships will be determined and then additional elements of knowledge will be added. In the process, differences in the conceptualizations of the six authors will be examined and resolved as much as possible.

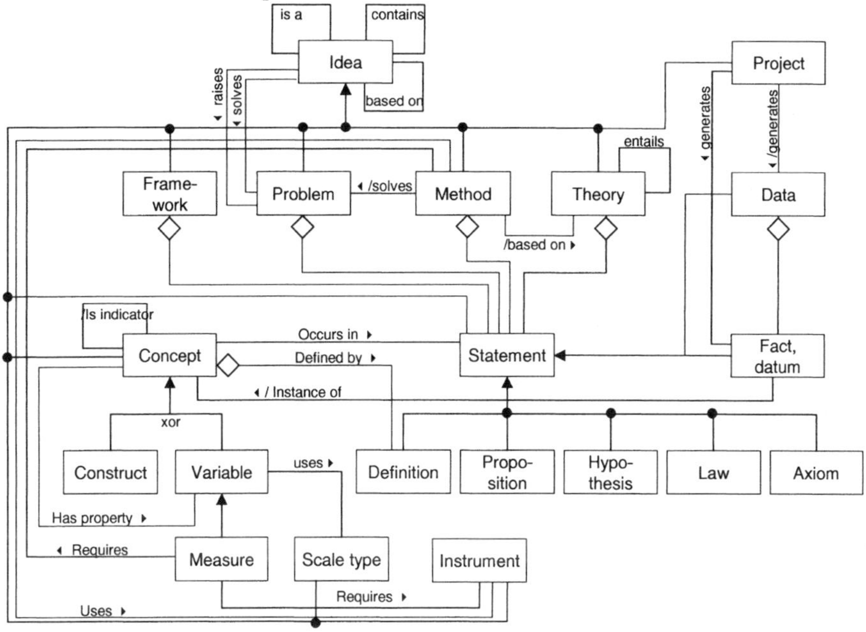

Figure 18. Conceptual model of scientific knowledge

All models contain similar core elements of scientific knowledge. Each model contains classes to represent scientific theories, statements and concepts. Theories are present in each epistemology, although the way that they are linked to other elements of scientific knowledge differs considerably. As a consequence, a 'theory' class needs to be part of the model of scientific knowledge. All four epistemologies contain 'statements' although Bunge prefers the term 'formula'. Nagel and Popper distinguish statements and statement-forms (i.e. statements which include variables). In the following the term 'statement' will be used to cover both interpretations. Dubin distinguishes four types of statements ('law of interaction', boundary criterion', 'proposition' and 'hypothesis'), although he does not explicitly generalize them into a separate 'statement' class. Thus a statement class needs to be added to the model. Another element of knowledge which appears in all four epistemologies is 'concept'. Bunge and Popper use this term while Nagel mostly speaks of 'terms'. Dubin adopts the term 'unit' to reduce the ambiguity of the term 'concept'. Thus scientific 'concepts' are present in all four epistemologies and need to be added to the model of scientific knowledge.

In all four epistemologies the interdependencies between statements, theories and concepts are similar. The authors aggregate statements into theories and agree that statements (or formulae) contain scientific concepts. Nagel and Bunge introduce intermediary terms between statements and concepts (e.g. 'term' and 'factual hypothesis') which can be subsumed under the term 'statement' in the model of scientific knowledge. Dubin distinguishes subtypes of 'statement' which will be considered further below. The core component of a model of scientific knowledge therefore consists of theories, which contain statements, which in turn contain concepts.

The epistemologies differ in the way that they refine this structure. All authors distinguish different types of statements. Popper, for examples, introduces 'hypothesis' and 'law' as special kinds of statements. Nagel and Bunge include similar terms. Dubin treats hypotheses as determinate empirical statements which do not contain theoretical terms and opposes them to propositions which may contain theoretical terms. Bunge and Popper do not place this restriction and use the term 'proposition' in its logical sense. However, Dubin does not provide a term which would correspond to 'law'. His term 'law of interaction' is closer to the general term 'statement' and is not intended as counterweight to 'theories'. Several authors distinguish between singular and universal statements. This distinction is important for introducing facts – which are singular statements – as another subtype of statement. Two additional subtypes of statements need to be added: definitions and axioms. They are present in the epistemologies of all authors. Definition statements need to be linked to the concept which is defined. This requires the introduction of a 'defined by' relationship between 'concept' and 'definition'.

The authors distinguish additional elements and relationships. Nagel, for example, discusses the phenomenon that one theory may be derived from another. He calls the process the 'reduction' of theories. Bunge extends this idea by introducing the concept of 'depth' of a theory. A theory is deeper than another theory if it entails the other theory, if it contains higher-level theoretical concepts and if it introduces a mechanism. Deeper theories are more specific, have higher empirical commitment and are better testable (Bunge 1998a, pp 575-585). The core aspect in both cases is the fact that one theory can be derived from another; in the synthesized model this will be called the 'entails' relationship.

Bunge generalizes concepts and theories into 'scientific ideas'. Therefore, an additional 'ideas' class is added to the model. The class will be useful as a generalization for additional elements of scientific knowledge which will be introduced below. Popper adds 'methodology' as a special type of theory. Methodologies consist of 'methods'. Therefore 'method' is added as an additional class. In practice, methods and theories may differ significantly. Therefore methods are treated as specializations of scientific ideas rather than as specializations of theories. Both methods and theories are aggregates of scientific statements. In addition, methods can be based on a theory. This requires the introduction of a relationship between the two classes.

An additional relationship proposed by several authors is the generalization relationship. Nagel, for example, introduces a generalization relationship between statements. Bunge identifies the same relationship for problems, for data and for theories. It can be applied for concepts, problems, theories etc. and therefore needs to be added to the 'idea' entity.

The epistemologies contain various subtypes of 'theory', 'concept' or 'statement' which need not be represented explicitly. It is not necessary, for example, to introduce separate classes for universal and singular statements. The same applies to Popper's singular and universal concept. In addition, Dubin's Law of Interaction, Boundary Criterion and System State can be represented with the current structure of the model. Popper's 'basic statement' is a special kind of statement. It does not need to be modeled as a separate class. The same applies to Nagel's descriptive and theoretical terms.

Dubin and Bunge, however, add two refinements to concepts. Bunge introduces 'variables' while Dubin introduces 'empirical indicators'. Additional subtypes of concepts such as measure and construct can be derived from the texts on research methodology. As these different subcategories play an important role in the practice of science, they need to be added to the model. Concepts can thus be broken down into constructs and variables. Variables can assume a range of different values; this requires the introduction of scales. A measure is a special type of a variable. The 'is indicator' relationship shows how one concept is an indicator for another. The forward slash in the name of the relationship denotes that this relationship can be derived from other relationships and attributes in the model, in this case the statement class. It allows to precisely describe this relationship between both classes. Measures need to be linked to a method for data collection. This is achieved through the 'requires' relationship between 'measure' and 'method'. This process may require an 'instrument'.

Bunge introduces another class: scientific problems. They are important for the process of knowledge creation and for the allocation of scarce research resources. Therefore they need to be added to the model of scientific knowledge. Like theories, methods, and concepts, they are subtypes of scientific ideas. Additional relationships are needed to show the relationship between problems and scientific ideas. Scientific ideas such as theories and concepts may give rise to a new scientific problem. At the same time other ideas may solve a problem. A strong link exists between methods and problems; the (derived) 'solves' relationship indicates which methods can be applied to solve a practical problem.

Bunge and Nagel include categories such as 'occurrence', 'event' and 'fact (in the sense according to Bunge)' which directly reflect empirical phenomena. These categories are needed for epistemological argumentation. However, they cannot be represented directly in an information system because they refer to the phenomenon itself rather than to its description. A knowledge infrastructure can only contain descriptions of phenomena. The model of scientific knowledge already contains a 'fact' category for capturing descriptions of singular phenomena. Therefore no additional class needs to be added.

The analysis of textbooks in research methodology yielded several additional categories. In the early phases of research, scientists organize knowledge in frameworks. As frameworks differ from theories and concepts, they need to be represented as a separate class. Another class is needed for the aggregation of facts into data. For practical considerations, a class needs to be introduced which represents individual research studies and projects.

The resulting model is shown in Figure 18. In the next section, the characteristics of these elements will be analyzed in detail.

3.4. Elements of scientific knowledge

The conceptual model developed in the previous section has identified core elements of scientific knowledge. Although most of these elements (e.g. concepts, theories, methods, frameworks etc.) are present in current scientific publications, they are usually not identified explicitly. As a consequence the way in which these elements are described differs significantly from author to author. It is difficult to ensure consistency between different entries and to reuse these units of knowledge across articles. In an online environment, little additional effort is needed to mark the elements of knowledge and to provide services which are specific for each element of knowledge. As one example, an electronic environment, can provide a functionality which allows scientists to browse the many competing definitions of a concept and to highlight the differences. This should increase the reuse of existing knowledge.

The purpose of this section is to examine the different elements of knowledge and to identify their unique characteristics. These characteristics are needed for storing the different kinds of scientific knowledge in an online knowledge infrastructure. In addition, they can be used as a template which scientists fill when adding knowledge of this type.

In this section, each of the elements of knowledge will be examined. Their attributes and relationships will be identified. In addition, the main activities in which these elements of knowledge are involved will be analyzed to identify potentials for information systems support. Figure 19 shows a summary of the attributes which will be identified in the following. It shows all attributes of the 'unit of knowledge' class as well as the additional attributes of its subclasses.

3.4.1. Concept

Concepts are the most basic component of scientific knowledge. All scientific knowledge involves concepts in one way or another. Scientific statements consist of concepts and concepts are "basic building blocks of theory." (Turner 1986, p 5) Empirical observations are described in terms of established concepts. In the physical sciences, examples for concepts are 'energy' and 'temperature'. In the social science, examples are 'utility', 'user acceptance' and 'trust'.

Concepts are first and foremost linguistic units:

> "A concept is a bundle of meanings or characteristics associated with certain events, objects, conditions, situations and the like. We abstract such meanings from reality and use words as labels to designate them." (Emory 1984, p 23)
>
> "They are abstract elements representing classes of phenomena within the field of study." (Babbie 1994, p 49) A concept is a "mental construct which se-

lects and summarizes an aspect of the observable world for theoretical attention". Concepts are "only mental creations." (Babbie 1994, p 115)

Ideally, concepts are "images with exactly delineated properties that can always be recognized with absolute certainty" (Schlick 1974, p 20). In a knowledge infrastructure each concept can be represented as a separate unit of knowledge which has certain attributes and which is engaged in relationships with other elements of knowledge. In the next section, these attributes and relationships will be identified and discussed.

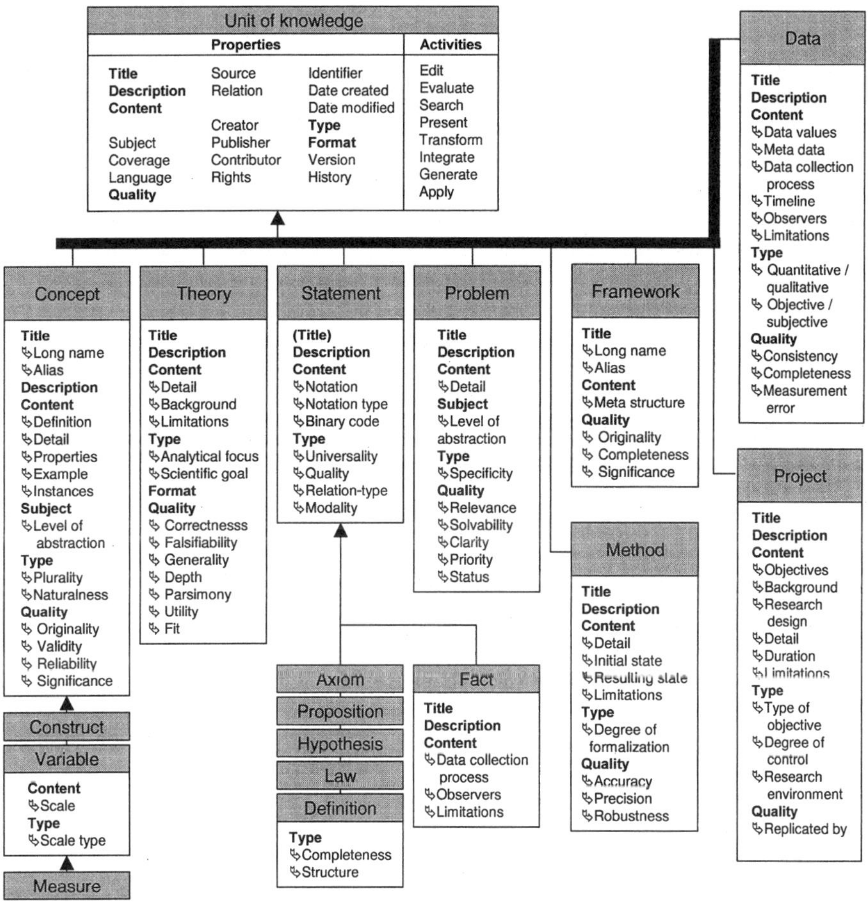

Figure 19. Characteristics of elements of knowledge

When concepts are stored explicitly in knowledge infrastructures, they need to be represented as bundles of properties, relationships and methods. Three different approaches can be utilized to identify these components. First, examples of current scientific concepts can be analyzed to identify characteristics of concepts. For instance, traditional scientific articles which deal with concepts such as 'user acceptance', 'knowledge', 'perfect competition' can be scrutinized. Properties can be identified by examining and comparing the meta structures of each concept description. An alternative approach is to analyze texts from the philosophy of science which describe key properties of scientific concepts. A third approach is to examine the uses of concepts and to infer the kinds of information which are necessary to enable such uses. In the following all three approaches will be utilized.

3.4.1.1. Attributes and relationships

The purpose of this section is to identify and describe the main attributes of scientific concepts which need to be represented explicitly in a scientific knowledge infrastructure. While the number of potential attributes is large and no exact boundary criterion can be specified, those attributes should be identified which are most valuable in scientific processes. The attributes which are identified in this section can be translated into corresponding database structures during implementation. Another implementation alternative is to use them for the creation of corresponding tags in a mark-up language such as XML. In either case, the identification of attributes allows the development of templates which allow scientists to describe the concepts that are central to their search efficiently and to reduce the chance for misinterpretations. In addition, they can be used to improve the effectiveness of search algorithms. They are also necessary for a wide variety of additional functionality for evaluation, consistency checking, integration, visualization etc.

An overview about the specific attributes of 'concept' is shown in Table 20. The table is structured into groups of generic attributes such as 'title', 'content', 'subject', 'type' and 'quality' which were identified as relevant for any unit of knowledge in section 3.3.2.1. above (see appropriately called 'presentation'. This leads to five core activities (create/edit, integrate, evaluate, search, present,) which will be further examined in the following. Two additional categories of activities are relevant: the generation of knowledge and its application (see Table 17).

on page 108). If a mark-up language is used for the implementation, then the groups become sections with additional subordinate tags for the 'concept'-specific attributes. If concepts are to be stored as structured records in a relational database, then all attributes need to be implemented as columns. Object-oriented databases have the advantage of being able to represent the relationship between generic and specific attributes more accurately.

3.4.1.1.1. Title

The most prominent attribute of a scientific concept is its name. Ideally, scientists should agree on the meaning of each scientific term and avoid using terms with multiple meanings. In scientific practice, however, this agreement is rarely achieved. In most cases, scientists disagree about the interpretation of a term. Many different interpretations exist, for example, for terms such as 'knowledge', 'information system', 'user acceptance' etc. Therefore a knowledge infrastructure needs to allow the coexistence of many concepts which carry the same name but which differ in their semantics.

In section 3.2.1., for example, three interpretations of 'knowledge' have been discussed: 'knowledge-that', 'knowledge-how' and 'knowledge by acquaintance'. In a knowledge infrastructure, each of these interpretations would be represented as a separate concept. Nevertheless, each author should be able to use the general term 'knowledge' in his text as long as he clearly indicates which of the variants he has in mind. This may include using the term 'knowledge' as sum of all existing interpretations of knowledge. Many current publications that focus on some aspect of 'knowledge' fail to identify the variant of knowledge which underlies their argument. This burdens the reader, increases misunderstandings and increases the risk that the author unknowingly mixes characteristics of several incompatible variants.

Table 20. Key attributes of 'concept'

Attribute	Description
Title	
↳Long name	Name which uniquely identifies the concept
↳Alias	Alternative term for referring to the concept (may overlap with other terms)
Content	
↳Definition	Link to definition statements (relationship)
↳Detail	Detailed description of the concept
↳Example	Description of a prototype plus examples on both sides of the concept boundary
↳Instances	The set of instances of this concept; or a method to enumerate the instances
Subject	
↳Level of abstraction	Theoretical/nominal vs. observational/real term
Type	
↳Plurality	Whether the concept represents a substance or discrete things
↳Naturalness	Whether the concept is a natural (basic) category
Quality	
↳Originality	Extent to which the concept differs from other concepts in the infrastructure
↳Validity	Adequacy of the meaning of the concept; absence of meaning-related measurement errors
↳Reliability	Requires consistent results for measurements
↳Significance	Relevance of the concept

As a consequence, a knowledge infrastructure needs to distinguish between two aspects of the 'title' attribute. On one hand, the title should allow to reference a concept with as little overhead as possible. On the other hand, it should allow to distinguish one concept from all others. Therefore the title attribute needs to be broken down into two attributes: one 'alias' attribute and one 'long name' attribute. The alias is a short form which can be used when the context is clear and when misinterpretations are unlikely. The 'long name' attribute uniquely identifies the concept and distinguishes it from all others. Examples for long names are 'knowledge-that', 'knowledge-how' etc. These names may be expanded when different interpretations for these subtypes of knowledge are proposed. Thus in a knowledge infrastructure, many alternative conceptualizations of knowledge might exist. Each of these concepts might carry the term 'knowledge' as alias. None of these concepts, however, would carry the term 'knowledge' as its title. By default unique concept titles could be constructed by including key aspects of an interpretation in the title or by appending the authors' name to the alias, e.g. 'knowledge according to XX'.

The distinction between a unique concept name (title) and concept alias may significantly reduce misunderstanding caused by incompatible interpretations of a term. Authors can clearly indicate which interpretation of a concept is relevant for their article. This may significantly reduce the problem of "researchers talking right past one another" (Stegmüller 1976, p 19), where each participant uses the same words with different meaning. Kuhn (1996, p 102) has pointed out that even in the physical sciences core concepts such as 'mass' have significantly shifted their meaning in the transition from Newtonian to Einsteinian Physics. Although scientists usually speak of 'mass' instead of 'Einsteinian mass' or 'Newtonian mass', these would be separate concepts in a knowledge infrastructure.

A direct consequence of this flexibility in naming will be an abundance of concepts with the same alias. Many variations of the terms 'knowledge', 'user acceptance', 'information system' may co-exist and these variations will participate differently in theories. Thus mechanisms are needed to reduce the variation in concepts by identifying commonalties between different concepts and by introducing generalizations which disregard variations in minor details. A knowledge infrastructure needs to incorporate mechanisms to identify, manage and resolve differences in the interpretation of a term.

3.4.1.1.2. Content

3.4.1.1.2.1. Definition

Another important aspect of a concept is its definition. The goal of a definition is to describe a concept precisely enough to distinguish it from all other concepts and to avoid "confusion about the meaning of the concepts." (Emory 1984, p 25) A definition, thus, establishes a clear boundary between the concept to be defined. Definitions do not provide insight about a concept beyond what is needed to establish the boundary. In scientific practice the goal of clear definitions is rarely reached. Articles often lack detailed, well-crafted definitions for their core concepts. Meaning is frequently established implicitly from the context in which terms are used and from the formulas in which they are embedded. This is not just an oversight. The problem of the hermeneutic circle fundamentally limits the ability to develop highly precise definitions.

Definitions are relational in structure. They relate one concept (the definiendum) to other concepts (the defeniens). A definition, relates a new concept to existing concepts. It can only be as sharp as the concepts on which it is based. The relational structure is represented in Figure 18 on page 121 by the 'defined by' arc between 'concept' and 'statement'. Definition statements are further described in section 3.4.2.2.2. beginning on page 151 below.

In an online environment, the role of definitions may change because their relational structure can be better exploited. The online environment allows instant access to definitions of the terms used in a definition. This is particularly useful if these terms have several possible interpretations (e.g. terms such as 'information system', 'knowledge' etc.). Beyond an increase in the clarity of definitions, the online environment has the added advantage that it may reduce duplications in definitions. In an online environment, all interpretations of a given term can be made available. Thus before an author introduces a new term, he can verify that no similar term has been proposed yet. This requires additional functionality for evaluating and comparing definitions.

3.4.1.1.2.2. Detail

It is not sufficient to describe concept boundaries in a definition. The 'detail' property is needed to capture a detailed description which includes structure, behavior, variants etc. Not all such knowledge needs to be explicitly stored in this attribute, however. It may also be embedded in the network of relationships which a concept is part of. 'Contains'-relationships, for example, may describe a concept's components. Theories in which the concept is involved may be used to describe its behavior and its variants.

3.4.1.1.2.3. Properties

This attribute describes the characteristics of a concept. The content (or value) of a property may vary from instance to instance. For example, the concept 'person' may exhibit properties such as 'income', 'education level' etc. Properties may themselves be concepts. Such properties are always variables or measures because they need to reflect specific information about each instance. Thus properties are relational constructs. They are represented in the model of scientific knowledge as a 'has property' relationship between concepts and variables. The term property used here is equivalent to the term 'attribute' used in object-oriented and data modeling methodologies (Norman 1996, pp 102-115).

3.4.1.1.2.4. Example

Because of the hermeneutic circle and because of the existence of different gradations of concepts rather than the existence of sharp boundaries a precise definition of a concept is rarely possible. One way to reduce the ambiguity is to provide examples. Prototypes of a concept are preferable, i.e. examples of the phenomena which are central to the category. Besides their ability to clarify meaning, examples can be used for initial tests of statements and theories involving the concept. Problems may become evident, when a concept is replaced by one of its instances (e.g. an example) in a statement or a theory. This can be done semi-automatically and is particularly useful for more complex theories where relationships are not obvious. One approach would be to generate sentences from a theory using the examples which could then be checked by a researcher for inconsistencies.

To further clarify the boundaries of a concept it has proven useful to add examples which are close to the boundary. This should include counterexamples – phenomena which are close to but outside of – the boundary. Counterexamples significantly increase the clarity of the concept. When a counterexample is provided, care should be taken that it is a case with potential for disagreement. This is an indication that it is close to the boundary. If an author is not capable of providing a good counterexample, this is an indication that the conceptualization is not yet sufficiently mature.

3.4.1.1.2.5. Instances

This attribute can be used to identify instances of the concept. In its simplest form the attribute contains a list of all instances. Most concepts, however, do not have finite sets of instances. An alternative is to provide a procedure for enumerating all instances at any time or location. For the concept 'enterprise', for example, such a procedure might be to extract the names of all companies from each countries' commercial registry. In the future, such procedures could be automated. The advantage of the instance attribute is that it enables other scientists to test a theory by randomly selecting elements to which the theory should apply.

3.4.1.1.3. Subject

3.4.1.1.3.1. Level of abstraction

In research practice, a distinction needs to be made about the level of abstraction of a concept. Nominal terms – which are also called 'theoretical' terms – need to be distinguished from real or observational terms:

> "With regards to an observational term, it must be possible, under suitable circumstances, to decide by means of direct observation whether the term does or does not apply to a given situation." ... "Statements which purport to describe readings of measuring instruments, changes in color or odor accompanying a chemical reaction, utterances made or other kind of overt behavior shown by a given subject under specified observable conditions - these all illustrate the use of intersubjectively applicable observational terms. Theoretical terms, on the other hand, usually purport to refer not to directly observable entities and their characteristics; they function"... "in scientific theories intended to explain empirical generalizations." (Hempel 1958, p 42)

Thus a theoretical concept is a purely abstract term while an observable term represents a clearly identifiable, 'real' phenomenon. As an example of a theoretical term, Hempel (1958, p 43) provides 'specific gravity', which is defined as quotient of weight and volume of an object. Specific gravity can be used to determine whether an object will float in a fluid. Weight and volume are observational terms on which this theoretical term depends. The dichotomy is not without problems because all observation is theory-laden. Bar-Hillel argued that "the distinction between theoretical and non-theoretical is the result of an arbitrary"... "slice in a continuum." (Bar-Hillel 1970, p 268) Indeed, it may be the continuum which matters (Babbitt 1997) and projection onto binary categories may be useful in specific contexts but not in general.

Hempel's example describes an ideal case: a nominal term is embedded in an equation in which all other terms are observational. The nominal term is fully determined by the observational terms. However, many nominal terms are not tied as closely to observational terms. The problem is especially relevant in the social sciences. Sztompka (1990, p 49), for example, adds a distinction between societal versus sociological non-observational concepts. During theory development researchers frequently introduce theoretical terms without strong empirical basis. Especially in the social sciences, nominal terms abound which are very difficult to translate into observational terms. Examples are 'trust' and 'intelligence'.

The distinction is important for the practice of science. Theoretical terms provide generality and can identify hidden relationships which are not obvious to the observer. They constitute progress in science. At the same time, theories need to translate back into operational terms. Researchers have few problems in devising new abstract theories. In addition, they are comfortable with collecting observational data and extracting empirical generalizations. However, the link between theoretical and observational concepts remains most challenging. Observational terms which can be linked to a theoretical term are called 'indicators'. For example, the observational term 'Standardized score on the progressive matrices test' is an indicator for the nominal term 'intelligence' (Miller & Wilson 1983, p 21).

In a knowledge infrastructure, the distinction between both types can be used for quality control. The maturity of a theory can be assessed by comparing the number of nominal and real concepts. Algorithms can determine which concepts are present in a theory and can then compute the mix of theoretical and nominal concepts. They can examine the relationships between the theoretical concepts and their indicators. Theories which have many theoretical constructs that lack indicators are under-constrained and are not likely to be falsifiable. Some adjustments may have to be made to take into account variations in the importance of the concepts. Algorithms may determine what concepts are most strongly in need for the development of indicators. If different indicators are used for a concept, discrepancies can be analyzed between the indicators. Similarly, the quality of a theory is higher when – ceteris paribus – already established indicators are used and the author of the theory does not need to create his own. The distinction can also be used to identify the research style of scientists: some scientists are more empirically inclined and thus focus on observational concepts while other researchers are more theoretically inclined and focus on theoretical concepts. This attribute can also be used to rank order concepts. This is relevant for assigning research priorities or for the visualization of concept spaces.

In the practice of social science research observational terms are favored over theoretical terms. Dubin, however, warns that in the process of research, a theory will often initially be based on nominal concepts which subsequently may be converted into real concepts. If researchers were discouraged from focusing on theoretical concepts that are not immediately tied to observational concepts, this might lead to the following problems:

> "(1) he probably will never attempt to discover new empirical indicators (as he cannot think of loooking for new things if his only tools of imagining are the representation of what he knows already), and (2) he probably will devote a majority of his research to wholly trivial problems." (Dubin 1978, p 51)

During theory development, it is often necessary to introduce new – and initially nominal – concepts. As research progresses, reliable indicators need to be found. The relative number of observational terms which form part of the theory, therefore, increases. This number can be used to measure research progress – although it may not be adequate to compare the state of knowledge between different theories. In summary, the distinction between nominal and real concepts - albeit only by degree - is important for theory building, testing and validation.

The distinction between nominal and observational terms has led to considerable debate about the relationship between a concept and the corresponding phenomena in reality. On one side of the argument are the instrumentalists who argue that it is sufficient to view concepts as suitable representations of some phenomena and that it is not necessary - and potentially impossible - to assume that a concept corresponds to some 'real object'. Miller and Wilson, for example, emphasize that concepts are unobservable and purely mental constructs (Miller & Wilson 1983, p 21 & p 112). Babbie states:

> "The terms associated with concepts are merely devices created for purposes of filing and communication. The word 'prejudice' is an example. Ultimately that word is only a collection of letters and has no intrinsic meaning." (Babbie 1994, p 115)

Proponents of the realist view counter that instrumentalists reduce scientific terms to no more than "conceptual links in rules of representation and inference" (Nagel 1961, p 140). They overlook the need for links to reality. Although the truth may lie in the middle between both positions, a scientific knowledge infrastructure must adopt the instrumentalist view. In scientific communication, terms cannot be taken for granted because of their alleged correspondence to some real phenomenon and misinterpretations will always occur. An instrumentalist knowledge infrastructure is able to accommodate the realist's ideal case that terms have a single 'real' meaning because of their correspondence to real phenomena. The converse is not true; a realist infrastructure would not be capable of handling the many conflicting interpretations of the same real phenomenon.

3.4.1.1.4. Type

3.4.1.1.4.1. Plurality

Concepts can be distinguished according to the way that their instances can be counted and aggregated. An instance of a substance can be divided and still remain an instance of the substance. Half a liter of water, for example, remains water; half a brick is still a brick. A little less user acceptance is still user acceptance. The process of division may have limits; once a certain boundary is reached, smaller components may not be regarded as instances of the concept any more. Instances of other concepts, may be discrete objects. Collections of these objects only exist in integer numbers. One can own one house, two houses, three houses.. etc. Values in between are not possible. Enumerative types are important because they determine what kind of set operations are possible on instances of the concept. Another distinction can be made among discrete concepts. Instances of concepts which may contain other instances of the same concept are called recursive. A house, for example, cannot consist of several houses. A database, however, may be an aggregation of several databases. A holding company consists of several companies, some of which may also be holdings.

3.4.1.1.4.2. Naturalness

As a consequence of prototype theory, the cognitively more salient categories need to be identified. Within a hierarchy of concepts, basic level concepts "maximize the number of attributes shared by members of the category"... and ... "minimize the number of attributes shared with members of other categories." (Taylor 1995, p 31) 'Chair' is an example of a basic category while furniture and kitchen chair are not. If a theory contains concepts which are more general than the corresponding basic categories, then they need to be carefully tested with all their basic categories. If a theory incorporates concepts which are below the basic level, an attempt should be made to generalize to the basic level.

3.4.1.1.5. Quality

In the literature, a large number of criteria have been proposed for evaluating concepts. In the following, these criteria will be briefly described. In addition, ways in which a knowledge infrastructure can support the evaluation process will be identified. The approaches range from gathering the assessments of reviewers and guiding reviewers through an interactive evaluation process to automatically analyzing the properties of concepts and the network of relationships in which a concept is embedded.

3.4.1.1.5.1. Originality

A concept needs to differ from concepts which already have been established. Scientists long have argued that every concept should be tested to see whether it makes a "sensible difference" (James 1911, p 189). In current scientific practice such tests are difficult because concepts are deeply embedded in the particular thought structures of individual manuscripts. In addition, manuscripts often lack the space for an explicit, detailed description of even core concepts.

Information technology, in contrast, can help to determine the originality of a concept by analyzing it for similarities with already existing concepts. In computer-linguistics, for example, much progress has been made on algorithms for natural language analysis and for automated text classification. (Lewis & Jones 1996; DARPA 1991, 1992) Linguistic synonym detection algorithms can be utilized to spot similarities. For an overview and examples of such methods see Hars (1998), and (1994, pp 180-194). The opportunities for IT-based analysis of originality increase further when concepts are embedded in a formal meta structure of scientific knowledge. This allows searching the set of already defined concepts for concepts which have similar explicit properties or relationships. Concepts may be similar, for example, if they occur in variants of the same theories or if they are close to one another in a generalization hierarchy. Unless the semantics of a concept are fully specified within a knowledge infrastructure, algorithms for detecting similarity cannot operate in full autonomy. The final decision whether a concept needs to be considered a duplicate of another concept has to be taken by a person who is familiar with the concepts.

An ideal situation to check for originality occurs when an author adds new knowledge. As part of the integration process, an author should be encouraged to reuse existing concepts. A knowledge infrastructure should automatically attempt to detect similarities between any newly introduced concepts and existing concepts using approaches described above. When overlap is detected, the author should be encouraged either to replace his duplicate concept with the existing concept or to clarify the differences between his concept and the already existing similar concept. This leads to better descriptions of concepts and reduces the potential for misunderstanding. The issues which occur during integration will be further discussed in section 3.4.1.3.4. below.

A second aspect of originality is the distinction between axiomatic concepts that cannot be derived from other concepts and derived concepts. 'Specific gravity' is an example for the latter because it is defined as the quotient of 'weight' and 'volume'. Axiomatic concepts are important because they are the most basic elements to which insights eventually need to be reduced. Derived concepts can be useful to simplify calculation and reduce reasoning complexity. A knowledge infrastructure can provide automatic algorithms which determine which concepts are axiomatic and which are derived. It is possible that several alternative sets of axiomatic concepts can exist in parallel for a domain.

3.4.1.1.5.2. Validity

Validity is one of the key criteria for assessing scientific concepts. It relates to errors that are associated with the interpretation and meaning of scientific concepts. Many different aspects of validity have been identified which scientists need to take into account when establishing concepts (Emory 1985, p 94). They will be discussed subsequently. A knowledge infrastructure can support the evaluation of validity by providing templates and guidelines for evaluation and by providing automated evaluation algorithms. Validity is a summary criterion; a concept is valid only if it meets all validity criteria.

Face validity: A concept meets this criterion if it "jibe[s] with our common agreements and our individual mental images associated with the concept." (Babbie 1994, p 127) From an interpretivist or realist perspective face validity depends on the inter-subjective nature of a concept. Face validity applies to theoretical as well as observational concepts. It does not restrict the ability of the researcher to define concepts but prevents him from assigning labels to concepts which induce false interpretations. If a positivist perspective is adopted which assumes the existence of an objective world with separate, identifiable objects, then face validity goes further: It requires that the concept provides an accurate representation of the real object. Face validity cannot be determined automatically. As natural language processing advances, it may become possible to identify some inconsistencies between accepted terminology and the characteristics that a researcher ascribes to a concept. A knowledge infrastructure needs to elicit the assessment of face validity from reviewers and readers and needs to support discourse when disagreement about the face validity of a concept emerges.

Construct validity: Scientific concepts are usually related to one another through theories and empirical generalizations. A measure should be faithful to the role of the concept in accepted theories or empirical generalizations. Should a conflict occur, then some change must be made to the interpretation of the concept and to the set of related theories. A measure for tacit knowledge, for example, should not conflict with empirical generalizations such as that learning and experience increase knowledge. Construct validity can be determined by examining the role of the concept in other theories. While information technology cannot determine construct validity automatically, it can support the reviewer by identifying other theories in which the concept is relevant and by displaying the role which the concept plays in these theories. Therefore information technology allows a more effective, interactive assessment of construct validity.

Content validity: This criterion "refers to the degree to which a measure covers the range of meanings included within the concept." (Babbie 1994, p 128) A measure which only reflects one of several aspects of a concept, therefore has low content validity. Conversely, 'salary' as a measure for 'income' has low content validity because other variants of income (e.g. interest income) exist which it does not cover. Content validity can in part be determined automatically. An algorithm can analyze the network of relationships between concepts, variables and measures to determine similar concepts. It can then identify omissions and thus flag potential problems in content validity.

Discriminant validity: This criterion applies to concepts which have been operationalized. It requires that such concepts "must be empirically distinguishable from one another." (Bacharach 1989, p 503) When empirical data is available for similar concepts, then some types of discriminant validity may be determined automatically. In more complicated cases, interactive statistical analysis may be required. Another alternative for supporting the evaluation in a knowledge infrastructure is to provide an opportunity for reviewers or readers to express doubts about discriminant validity. These doubts need to be recorded because they may indicate new or previously not considered relationships between concepts. Discriminant validity can also be analyzed on the level of concepts. It requires the absence of different concepts which share the same meaning. Therefore algorithms for the detection of synonyms and overlap between concepts are also relevant for ensuring discriminant validity.

Convergent validity: This criterion is used when several alternative measures exist for the same concept. In such a case, these measurements must share variance (Schwab 1980). Similarly to discriminant validity, this criterion can be derived automatically from the network of concepts provided that empirical data is available.

Criterion-related validity: Criterion related validity can be applied if a measure is used for "some empirical estimating purpose" (Emory 1985, p 95). The validity of a driver's test (a measure) can be determined by examining how well the drivers perform (Babbie 1994, p 127). Another example of criterion-related validity is "an opinion questionnaire which correctly forecasts the outcome of a union election" (Emory 1985, p 95). Upon closer examination, this form of validity, is highly similar to construct validity: it is based on the way that a concept is related to other concepts. A driver's test is not a direct measure of how well drivers perform. It is a measure for a theoretical concept which might be called 'driving ability'. The latter may be part of a theory of driving performance. Thus the measure can be evaluated by comparing the actual relationship between driving ability and driving performance with the hypothesized relationship. A similar issue arises in the election case: The questionnaire is not a direct measure of voter behavior; at best it can measure voter intent. This may be highly predictive by itself or it may require some transformations to arrive at voter behavior. In both, cases voter intent and voter behavior are related through a theory. Criterion-related validity therefore is based on relationships between concepts which are based on theories or empirical generalizations. Similar to construct validity, IT therefore allows interactive evaluation which may refine the network of concepts and theories.

3.4.1.1.5.3. Reliability

This criterion is typically applied to measures. A measure is reliable "if it produces consistent results."... "Reliability is concerned with estimates of the degree to which measurement is free of random or unstable error." (Emory 1985, p 98) Reliability can be divided into two criteria: Stability and equivalence. A measurement is stable, if the same measurement, taken at a different time, yields the same result. Stability is difficult to achieve if situational factors can affect the observation. A second aspect of assessing reliability is equivalence. A measure is high in equivalence, if different observers rate the same events in the same way. When a measure consists of similar items, then a respondent should rate them consistently. A variety of measures – for example Cronbach's Alpha – have been developed to assess the reliability of a measure. Reliability can be computed when empirical data is available.

Although 'reliability' is typically not applied to concepts other than measures, a similar problem arises for constructs and variables. If different persons attach significantly different interpretations to a concept, this is an indication of a deficit in its formulation. The difference between both conceptualizations of reliability is that it can be objectively determined for measures while it remains subjective on the level of constructs. On the construct level, reliability is closely related with clarity and the ability of a person to form a 'mental picture' of the concept.

3.4.1.1.5.4. Significance

Significance or relevance, rather than truth may be the ultimate criterion for judging a concept:

> "A concept is neither true nor false, only propositions are. A concept is neither valid nor invalid, only arguments are. Yet there is a distinction of "good" and "bad" among defined descriptive concepts. To have a name of it I shall say that a concept either is or is not *significant.* A concept is significant if and only if it occurs, together with others, in statements of lawfulness which we have reason to believe are true. It follows that some concepts are, in an inherently vague sense that cannot and need not be made more precise, "more" significant than others. For instance, a concept that occurs only in one or two tentative and isolated laws is "less" significant than one that occurs in a well-established theory of considerable scope. It follows, furthermore, that what is not significant today may become so tomorrow." (Bergmann 1957, p 50)

Thus the more strongly a concept is embedded in a network of concepts, the higher is its significance. A concept is preferable over another if it is relevant in more theories, statements or problems. If knowledge is represented within the meta structure, an indicator for significance can be computed automatically: An algorithm can determine the number of statements in which the concept occurs via the 'used-in' relationship. Statements, in turn, are related to theories. Thus for each concept, the number of related theories can be determined. This number can be used for reducing the search space during browsing and to identify the most promising concepts within a group of related concepts. In addition it can be used to judge a contribution. A concept ultimately is more valuable if it is related to a larger number of existing concepts and theories. An automated measure for significance has the drawback that paradigm-changing concepts might initially have low significance. This problem may be overcome if novel concepts are not evaluated on significance alone but also on their degree of originality.

3.4.1.1.6. Relationships

Concepts are related to most other elements of knowledge. In the model of scientific knowledge, most of these relationships occur indirectly via statements. Statements are linked to concepts via the 'contains' relationship. Statements, in turn contain other concepts and are part of theories, methods etc. By analyzing these dependencies, a knowledge infrastructure can provide reports about the sets of concepts which are relevant for theories and can compare overlap between theories. Similarly to their indirect relationship to theories via statements, concepts are related to methods, problems and frameworks. Moreover, concepts are related to facts. The model of scientific knowledge shows a derived 'instance-of' relationship between them. It relates concepts to facts describing their occurrence in the real world. Besides being related to other elements of knowledge, concepts are linked to one another via generalization, part-of and has-property relationships. It is a major challenge for academic knowledge infrastructures to develop interactive editing software which presents and manages this web of relationships.

Although concepts and theories are modeled as different classes to be linked via relationships, the boundaries are fuzzy. A sharp definition of a concept requires a detailed description of its characteristics and interdependencies with other concepts. The more precise these statements are, the more similar they may become to a theory about the concept. A precise characterization of the concept of 'knowledge' for example, quickly becomes a theory of knowledge. However, such close relationships between a theory and a concept do not always exist and they can be automatically detected by analyzing the network of relationships.

3.4.1.2. Subtypes

In research practice, different types of concepts are distinguished: constructs, variables, measures and indicators. The characteristics of these subtypes will be examined in the following. Constructs, variables and measures are positioned on decreasing levels of abstraction. An example of such a hierarchy is shown in Figure 21 on page 141. It will be discussed in the section on 'measures' below. Concepts are often involved in networks of relationships where one concept is used as an indicator of another concept. This is reflected in the 'is indicator' relationship between concepts.

3.4.1.2.1. Construct

In the early phases of research, scientists frequently work with concepts which are vague, incomplete and which have no clearly established boundary because they don't directly relate to empirical phenomena. Such unobservable concepts are called 'constructs' (Gibbs 1972; Babbie 1994, p 113). An example is 'social class' (Emory 1984, p 25). Constructs are purely nominal or theoretical concepts. During theory formulation, constructs are replaced with more limited but clearly defined variables. The 'social class' construct, for example, may be refined into a 'social class' variable which is based on specific combinations of income and education levels. This may be only one of many alternative 'social class' variables. The transition from construct to variable is always problematic because it involves a change in meaning.

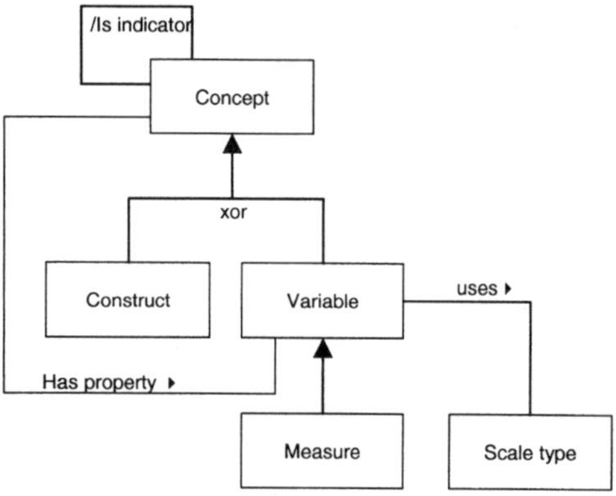

Figure 20. Subclasses of 'concept'

Compared to the general 'concept' category, constructs do not introduce additional characteristics. The characteristics of concepts outlined above suffice for their description. Similar to concepts, the larger the number of theories which are related to a construct, the higher is its significance. Especially in the social sciences, researchers need to avoid the creation of ad-hoc constructs.

3.4.1.2.2. Variable

Variables provide the foundation for empirical research. In contrast to constructs, variables are observational concepts. They categorize a range of observable phenomena. Each permissible value of a phenomenon is called an attribute. An example of a variable is 'gender'. It consists of the two attributes 'male' and 'female'. An alternative term for variable is 'unit' (Dubin 1978).

Variables have two additional attributes: The scale used for describing the range of values which a variable can adopt and the type of the scale. Values may consist of integer, real or other sets of numbers. Scales may also consist of non-numeric linguistic terms such as 'excellent', 'good', 'fair'. Variables differ in the type of scale used and in the relationship between the values of a scale (see Table 21). A variable uses a 'nominal' scale if its attributes form an unordered set of discrete values. Examples are 'gender' or 'marital status'. The set of values needs to be defined explicitly e.g. {male, female} or {single, married, divorced, widow}. Sets of values need to be exhaustive; it must be possible to map every instance of the phenomenon to a value and values may not overlap.

Nominal variables occur frequently in social science research. Because of the absence of an ordering criterion, they are not well suited for use in mathematical equations. To embed a nominal variable in a set of equations, existence equations need to be added for each attribute where the existence of the attribute is represented with a value of '1' and its absence with a value of '0'. A more suitable approach is the use of predicate logic. It is more flexible in treating issues of existence and conditional relationships between values of attributes.

Table 21. Types of scales

Type of scale	Characteristics of scale	Basic empirical operation	Example*
Nominal	No order, distance or origin	Determination of existence, equality	Gender Marital status
Ordinal	Order, but no distance or origin	Determination of greater or lesser values	Prejudice Alienation
Interval	Both order and distance, but no origin	Determination of equality of intervals or differences	Intelligence Temperature in Celsius
Ratio	Order, distance and unique origin	Determination of equality of ratios	Income, age Temperature in Kelvin

Adapted from Emory (1985, p 87). For further refinement see Chmielewicz (1994, p 67).
* Babbie 1994, pp 135-137.

A variable uses an ordinal scale if the values of its attribute are discrete and can be logically rank-ordered. Examples are 'alienation' or 'prejudice' (Babbie 1994, p 135). If values are ordered and distance between values is meaningful, then the variable uses an interval scale. If values are ordered and distances as well as ratios are meaningful - which implies the existence of an origin or 'zero' point - then the variable is based on a ratio scale.

3.4.1.2.3. Measure

Measures are the most concrete subtypes of 'concepts'. A measure operationalizes a variable (Bacharach 1989, p 498). It refers to an observational term for which at least one range of values has been defined. In addition, it requires a procedure for mapping observations into the range of values. If observation occurs in the context of an experiment, then the description needs to include the setup and execution of the experiment.

Figure 21 illustrates the distinctions between the three subtypes of concept. It describes some constructs, variables and measures related to 'temperature'. On the top of the hierarchy is a generic 'temperature' construct. This is the common denominator for all meanings 'temperature'. It captures intuitive notions about temperature and thereby constrains the range of possible interpretations of the concept. This level, contains linguistic and observational statements such as that 'heat' and 'cold' are states of temperature, that temperature changes may transform fluids into gases, that fire increases the temperature, that it is colder in winter than in summer, that substances which are in contact with each other adjust their temperature to a common level etc. Thus the top-level construct captures the collection of interpretations and observations associated with temperature. The next level contains several variables and one construct. The generally accepted interpretation of 'temperature' in the physical sciences is as 'physical temperature' which is defined as "mean kinetic energy of molecules." (Bavink 1932, p. 99 cited by Nagel 1961, p 341) It uses a ratio scale developed by Kelvin. Many measures exist for this variable. They range from using the readings of a thermometer to analyzing infrared emission. Historically, other variables for temperature were developed by Celsius, Fahrenheit and others. They led to slightly different equations and did not incorporate a characteristic of temperature that was not known in Celsius' and Fahrenheit's' time: the existence of a minimum temperature - absolute zero. Kelvin's discovery allowed the creation of a ratio scale. His variable was superior than those of his competitors. It allowed to rewrite the body of equations relating to temperature in a uniform notation. At the same time, it was possible to maintain backward compatibility with the other historical variables (Celsius, Fahrenheit etc.) by extending their ranges to absolute zero and then defining conversion functions.

The interpretation of temperature used in the natural sciences, however, is not the only possible. Some of its insights violate our sense-impressions: thermometer readings sometimes conflict with our experience of 'hot' and 'cold'. Figure 21 therefore contains another construct 'subjective temperature'. This construct captures intuitions about subjective temperature; it does not yet prescribe what scale should be used. While the physical scales are obvious choices other alternatives such as 'cold', 'warm', 'hot' are alternatives. These scales might be important to determine whether different people have different perceptions about temperature. Examples for subjective variables are shown: perceived temperature in Kelvin can be based on the temperature reported by subjects in experiments. It can be compared to a variable which accounts for the chilling impact of wind on the skin of a subject.

This example of constructs, variables and measures shows that at any given time, the 'same' concept may occur in different instantiations within the scientific body of knowledge. Concepts may overlap and compete. But the significance of a concept can be assessed by the number of equations or theories in which it occurs. This shows, for example, that the 'temperature corrected by windchill factor' or 'subjective temperature' are less significant than 'physical temperature (ISO/Kelvin)'. (For a historic analysis of theories of heat see: Goldstein & Goldstein 1978, pp 63-113).

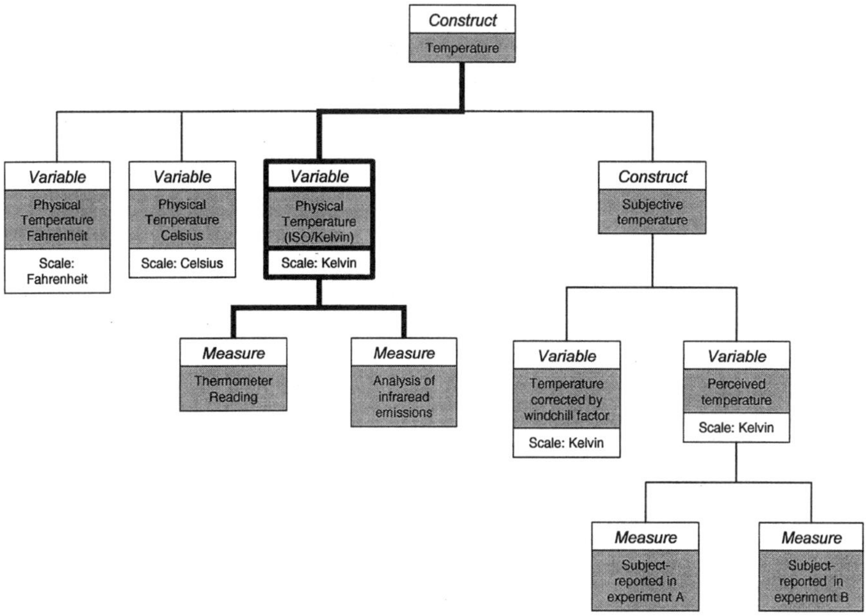

Figure 21. Hierarchy of constructs-variables-measures for 'temperature'

3.4.1.3. Activities

3.4.1.3.1. Editing

Concepts are described in natural language. Therefore standard text editors are sufficient. They may be combined with a tool for drawing concept maps (see below in the section on presentation).

3.4.1.3.2. Evaluation

Detection of inconsistencies: The meta-structure allows the definition of a variety of relationships between concepts. This includes generalization relationships, attributes and indirect relationships that can be derived from participation in statements. These relationships can be used for checking the consistency of concepts and for detecting potential problems. For example, if concepts share the same attributes and are used in the same theories, then they might be synonyms. If several attributes usually occur together, then the description of a concept which only contains a subset of these attributes might be incomplete. With the exception of obvious cases such as cycles in a generalization hierarchy, these algorithms only simplify the search for inconsistencies by identifying likely candidates. The final decision remains the task of a human. Nevertheless, such algorithms can significantly improve the quality of knowledge by identifying errors which would go unnoticed in traditional articles.

Template-based interactive critique: The knowledge which is embedded in already defined concepts can be used to systematically critique or examine new definitions of concepts. A knowledge infrastructure, for example, may generate questions to test whether a concept has been placed in the right position within a generalization hierarchy. This may be especially useful when concepts have been linked to observations. The system can identify categories which are conceptually close to one another in the generalization hierarchy and then generate questions to determine whether the boundaries have been placed accurately. In addition, the system can use an analysis of attributes and relationships to identify related concepts. If these concepts are not synonyms, they might be useful for associative reasoning. The knowledge stock accumulated in a knowledge infrastructure can be used to simplify the generation and validation of new knowledge.

3.4.1.3.3. Presentation

Concept maps: Scientific concepts are highly interdependent. They are embedded in a complex network of relationships. Advances in research continually change this network. Dependencies between concepts change as definitions are revised, as new concepts are introduced and as new theories are proposed. In the paper-based environment it has not been possible to systematically document and visualize these interdependencies because of the inability to integrate and update knowledge. This problem no longer persists in IT-based knowledge infrastructures. Interdependencies between concepts, statements, theories etc. can be captured in a repository which is based on the model of scientific knowledge described above. Algorithms can identify similarities between different concepts based on their characteristics, their common relationships or their occurrence in the same theories and methods.

This raises the problem of adequate approaches for presenting dependencies between concepts to the user. In the field of information systems, several methods for presenting relationships between concepts in diagrammatic form are already in use. Examples are entity-relationship diagrams and object oriented diagrams. These diagrams show concepts with their attributes and relationships. Approaches exist for generating such diagrams automatically. Another alternative is the creation of concept trees. Since Aristotle, researchers have recognized the value of embedding a concept within a generalization tree. While many attempts to develop a single consistent generalization tree – or ontology – have failed, it remains desirable to identify the main relationships in which each concept is involved. Even if the consolidation of all generalization relationships leads to some inconsistencies and requires multiple generalizations and several roots, such trees clarify the meaning of concepts. Another visualization technique – however with less clearly defined semantics – are mind maps.

An adequate presentation mechanism enables the researcher to quickly understand dependencies between concepts. Researchers avoid missing concepts which are relevant for their research and introducing new concepts which duplicate already established concepts. A problem for visualization approaches is avoiding information overload when the number of relevant concepts is high. This can be achieved by presenting dependencies on different levels of granularity. It requires methods for determining the relevance of a concept depending on the context. For example, techniques from network analysis can be utilized to determine the centrality of a concept (Nelson & Mathews1991).

As a consequence, the user-interface of a knowledge infrastructure should be able to visualize concepts in different types of concept maps. In addition, it should be possible to graphically add and edit contents to the knowledge infrastructure.

3.4.1.3.4. Integration

Balancing of concepts: Researchers frequently disagree on the meaning of a concept and often introduce new variants of a term. The body of scientific knowledge abounds with many interpretations of core terms such as 'knowledge' or 'information system'. This results in frequent misunderstandings and reduces the ability to develop theories. The current environment lacks mechanisms for consolidating variations of a term. Review articles may attempt to distinguish the main variants of a term (for example 1) knowledge-that, 2) knowledge-how and 3) knowledge by acquaintance for the term 'knowledge' or information system 1) as technical computer system consisting of hardware/software, 2) as social system which processes information and involves computer technology or 3) as any social system which processes information). However, they may not achieve sufficient impact on the terminology used across a scientific community.

In an electronic environment, it is possible to treat different interpretations of the same term as different concepts. This enables every author to define a variant of a term in precisely the way that he thinks of it. In addition it enables all members of a scientific community to view the whole range of variants of a concept. With appropriate analysis functionality, it is possible to identify those variants of a concept which are more widely used in articles or as part of theories. This simplifies reuse of existing concepts for new authors. In addition, an explicit range of variants allows scientists to examine the differences and to engage in consolidation of variants. Over time, scientific discourse may thus lead to a small set of key variants of a concept without suppressing the ability of the researchers to add 'niche' interpretations of a concept as he sees fit. When concepts are embedded in a network of relationships and when characteristics of a concept are defined systematically, then it is also possible to automatically identify overlap and differences between variants of a term. Such an analysis may lead to the generation of additional variants through permutation of characteristics. Thus a concept network may be redefined over time by subsuming minor variants under newly identified major variants. It is difficult to overestimate the clarifying effect which the need to expose synonyms and homonyms and the need to distinguish broad concepts from scientific concepts will have on research.

The ability to select between alternative interpretations of a scientific concept does not imply that authors should be required to identify the correct variant for every term. This might be counterproductive. Scientific statements often need to incorporate ambiguous terms from colloquial language and no ultimate solid foundation for such terms can exist. In addition, scientists sometimes want to avoid using a single variant of a term and instead refer to the most general interpretation which includes the sum of all variants. In the following, such concepts will be called generic variants of a term. Generic concepts may be highly ambiguous; they do not imply a fixed and precise interpretation. This is useful for discourse even if the most general interpretation is not necessarily consistent. Wide interpretations of a concept are useful for defining research problems; generic concepts are useful while narrowing in on a scientific concept and they may help to establish context. The introduction of a generic variant has the advantage of introducing a common denominator for scientists using different interpretations of a term.

As a consequence, knowledge infrastructures cannot universally enforce precision in the use of concepts. In addition authors need to be able to indicate that they intentionally wish to refer to a wide interpretation of a concept. The ability to distinguish open, broad generic concepts from clearly defined scientific concepts allows communication about difficult concepts such as 'knowledge' and 'information technology'. It is important to establish common ground in the early phases of theory building. Generic concepts, however, are inappropriate for a formal theory.

Identification of synonyms and homonyms: A crucial task of a knowledge infrastructure is to avoid redundant entries. In natural language processing much progress has been made for identifying synonyms. Thesaurus-based approaches as well as approaches for lexical analysis can be utilized for synonym detection (Hars 1994, pp 180-193). Synonyms may also be identified by comparing properties and relationships of different concepts. An information system can only identify candidates for synonyms. The final decision whether two concepts need to be merged must remain with a human. The automatic identification of a homonym is more problematic because it requires separating out different bundles of relationships from a single concept.

3.4.1.3.5. Generation

When concepts are embedded in a network of concepts and concept variants, the differences between the concepts can be analyzed systematically. Properties and relationships can be clustered giving rise to novel concepts. For example, when several different concepts share the same properties and relationships it may be possible to generate a generalization relationship between them.

3.4.2. Statement

Scientific knowledge arises from the interplay of many components of knowledge. Statements integrate these elements by describing their interaction and relationships. To create a theory, for example, many statements, often in mathematical form, need to be combined. Each statement, in turn, is built from concepts which are related to one another in specific ways. Other statements belonging to a theory may reference additional knowledge elements such as methods, facts, etc. Because of the complexity of their relationship, statements need to be treated as a separate class in the model of scientific knowledge (see page 121 above). Only key relational aspects of a statement, such as which concepts appear in a statement, which theory the statement belongs to (if any) etc. can be captured via explicit relationships within the model of scientific knowledge. Much additional detail about relationships between different concept is captured in the inner form of the statement, which may vary greatly. Statements may take the form of mathematical equations, logical or procedural expressions, diagrams or natural language text. These different forms will be further described below.

The explicit relationships in which statements are engaged can be used to determine the importance of elements of scientific knowledge. Network analysis, for example, allows to determine elements that are central to a domain. Concepts, which occur in many theories – or, more precisely, which appear in many statements belonging to some theories - may be more important than concepts which are relevant in very few theories. If more data is available for a concept, i.e. if many statements exist which relate observations to the concept , then the concept may be more mature than other concepts.

An additional problem for representing statements in a knowledge infrastructure is how to represent their context. One approach would be to describe different contexts explicitly by introducing a 'context' class and to relate each statement to the contexts where it is applicable. A more effective approach is to restrict the context by extending a theory, problem or method with statements which reduce the range of some variables. This approach has the advantage that the introduction of an inherently ambiguous additional 'context' class can be avoided.

3.4.2.1. Attributes and relationships

This section describes the additional attributes and relationships which are specific for scientific statements. An overview is shown in Table 22. In contrast to concepts, the 'title' attribute is of little importance for statements because statements are usually identified through their context (i.e. through their relationships to concepts and theories) rather than as individual objects.

Table 22. Attributes of 'statement'

Characteristic	Description
Content	
↳ Notation	Specific notation used for describing the statement
↳ Notation type	Mathematical, Logical, Procedural, Graphical or Linguistic
↳ Binary code	Binary representation of the content
Type	
↳ Universality	Degree of universality (universal, particular, singular)
↳ Quality	Distinction between affirmative, negative and infinite statements
↳ Relation-type	e.g. Categorical, causal, part-of, generalization
↳ Modality	Assertoric, Problematic (possibility), Apodeictic (necessity)

3.4.2.1.1. Content

3.4.2.1.1.1. Notation & notation type

These attributes describe the notation used for the inner form of the statement. In the natural sciences, the preferred notation type are mathematical equations. In the social sciences, the predominant form used for scientific statements is natural language. In engineering, graphics and diagrams are widely used. Logic-based and procedural statements are less frequently used. The notations differ in the precision of their semantics. While mathematical, logical and procedural statements have well-defined semantics, the semantics of graphical statements are more problematic; statements in natural language lack well-defined semantics. Note that the notation is not the only factor which determines ambiguity: precise semantics of the inner structure of a statement cannot make up for ambiguity in the concepts which are part of the statement. The precision of the inner structure cannot make up for the ambiguity of the outer form. From a knowledge infrastructure perspective, the advantage of using notations with well-defined semantics is that computer-supported algorithms for consistency checking and integration can be developed more easily. However, notations with well-defined semantics are not suitable for all sciences. Furthermore, precise semantics of a notation have few advantages if the concepts used in a statement lack precision. In the following, each notation type will be briefly described. Each notation type can be broken down into different sub-categories or notations.

3.4.2.1.1.1.1. Mathematical statements: equations

Mathematical expressions are widely used in the sciences to formalize knowledge. They consist of mathematical symbols representing variables and widely accepted operators (e.g. '+,-,*,/'), relations (e.g. '=<>'), functions ('√,∑,∫'), constants ('∞, π, e') with clearly defined syntax and semantics. Standardized symbols are often combined with user-defined symbols. Mathematical expressions can be integrated into larger bodies of knowledge by identifying the overlap in symbols used. The higher the standardization of symbols, the easier is the identification of such overlap. In the natural sciences, a higher amount of standardization has been achieved than in the social sciences.

An example for a mathematical statement is the basic formula for determining the optimal lot size (x_0) (Gutenberg 1965, p 204):

$$x_0 = \sqrt{\frac{2 \cdot u \cdot a}{p}}$$

This equation establishes a relationship between the concepts represented by the variables u, a, p, and x_0. Thus the optimal lot size (x_0) depends on the fixed production costs for each lot (a), the storage costs per period (p) and the demand per period (u). The links between the statement and these variables is established by the 'used in' relationship shown in Figure 18. Statements add constraints on the behavior or value that instances of concepts can assume and describe relationships between different concepts.

3.4.2.1.1.1.2. Logical expressions

The goal of logic is to provide a foundation for reasoning. Many logical systems have been developed to support rational reasoning. The most widely used system is first order predicate logic which consists of predicates, quantifiers, designators and variables (Guttenplan 1997). In contrast to mathematics, logic is not primarily focused on quantitative relationships. An example of a logical expression is:

$$\forall x(\text{StandardProduct}(x) \wedge \text{PredictableDemand}(x) \Rightarrow \text{MakeToStock}(x)),$$

where the predicate StandardProduct yields true if x is a product which is highly standardized, the predicate PredictableDemand(x) is true if demand for the product varies little and the predicate MakeToStock is true if make to stock is an appropriate production strategy.

First order predicate logic has significant limitations and therefore many extensions have been developed. In the domain of management science, logic is currently not used widely to represent knowledge. Although several efforts are under way by computer scientists to develop large systems of knowledge which include knowledge about business (see section 2.2.6. on page 50), business researchers so far have not adopted logic as a language to formalize their theories. In top management journals such as Management Science, Organization Science, Academy of Management Journal, knowledge is rarely provided in the form of logical expressions.

3.4.2.1.1.1.3. Procedural

Knowledge is sometimes represented in procedural statements. An example is the specification of a lot-sizing algorithm shown in Figure 22.

Initialization

```
solve P(t+1)'
u:= t_max; compute δ_v (v∈{t_1...,u})
determine s; compute γ_1 (r∈{s_1..., t_max})
moved:= no
```

Iterations

```
for t:= t_max-1 down to t_min do
        take solution of P(t+1)'; move f units from t+1 to t
        if l_t<C, then u:= t
        if t<s then
                determine s; compute γ_1 (r∈{s1..., t})
                moved:='no'
        if moved='no' then (Step1:)
                move production if profitable → solution
                of P(t)'
                if production moved then
                        moved:='yes'
                        u:=t
                        update δ_τ(τ∈{γ_t,...,t})
        perform Step 2
```

Figure 22. Example a of lot-sizing algorithm
Source: Van Hoesel & Wagelmans 1996, p149.

Currently only few articles describe knowledge in procedural notation. Larger procedures are difficult to read for the human. They are very sensitive to printing and entry errors and they are only useful if the reader has access to the same execution environment as the author. An online environment, however, eliminates the need to manually enter the procedures and the author can often supply the execution environment together with the algorithm. Thus an online infrastructure allows immediate execution of a procedure and can provide support for interactive analysis of the procedure. In the future the share of scientific knowledge formulated in this type of statement may rise significantly.

3.4.2.1.1.1.4. Graphical

Although diagrammatic statements play a major role in engineering and some of the natural sciences, they are less important in the social sciences. In management science, graphical statements have become popular as an element of business process re-engineering and for the analysis of requirements for information systems (Olle et al. 1991; Scheer 1998a, 1998b; Schmidt 1996; Brinkkemper, Lyytinen & Welke 1996). In these areas diagrams are used to develop abstract representations of a business. An example which graphically describes an order handling process is shown in Figure 23. The diagram contains information about the sequence of activities, the events which trigger each activity and the events which are the result of each activity. In addition, it identifies the organization units which are responsible for an activity and the objects on or with which the activities are performed. Although such diagrams have a formal syntax and therefore can be translated easily into more formal languages (e.g. first order logic), their semantics are rarely fully defined. This is not an oversight on the part of the researchers who use graphical statements but a consequence of the complexity of the domain. In many cases the number of possible exceptions and variations to a realistic business process is prohibitively large. To keep a process model manageable, first its core features need to be described. As humans are able to reason effectively under ambiguity, the lack of precise diagram semantics is not necessarily a grave disadvantage. The fact that researchers present scientific knowledge in diagrammatic format with ambiguous semantics can be interpreted as evidence of shortcomings of first order logic for representing such information.

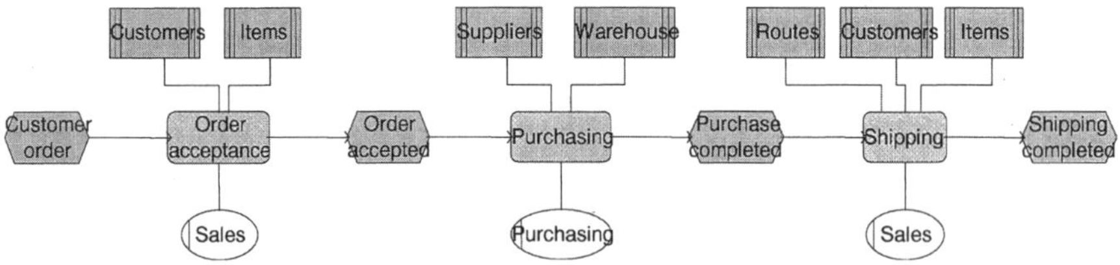

Figure 23. Example of a business process model
Source: Scheer 1998a, p 88.

3.4.2.1.1.1.5. Linguistic

In management science, much of the knowledge is presented in natural language. An example is the following statement which was used as the basis for a subsequent formalization of the concept of 'expected value of marketing expertise':

> "The value of marketing information is the difference between the expected value of decision outcomes with and without that information. The value of in-

> formation or expertise becomes the incremental expected benefit that expertise provides." (Pasa & Shugan 1996, p 372)

Such descriptions of knowledge in natural language are fraught with several problems: It is difficult to separate scientific language from everyday language; the inevitable combination of both and the ease with which meaning can be shifted increase the ambiguity of natural language sentences. In the statement quoted above, for example, the authors begin with the concept 'information', then equate 'information' with 'expertise' and finally continue with the term 'expertise', discarding the term 'information'.

In contrast to the paper environment, online infrastructures can provide features to reduce these problems. For example, they can extract the main terms used in a definition statement and look up the corresponding scientific concepts in the infrastructure's repository. They may then prompt the author to select his intended interpretation among several variants of the concept. This reduces misinterpretations and forces authors to be more precise in their statements. Other approaches are natural language processing algorithms for the detection of synonyms and homonyms, algorithms for information retrieval and for the categorization of natural language texts. Nevertheless, developing algorithms for quality control, summarization and the detection of inconsistencies in natural language texts remains a hard challenge.

3.4.2.1.1.2. Binary code

This attribute captures the raw content of a statement. It contains the binary code of the information reflected in the statement using the particular notation and notation type. This may be an equation in written format, a piece of source code for procedural notations, a diagram file or a text in HTML or ASCII encoding.

3.4.2.1.2. Types: Universality, quality, relation-type, modality

Different types of scientific statements can be distinguished. In the 18th century, Kant proposed a classification of statements which is still applicable today. He developed a system of four main categories of 'judgements', each of which was further divided into three subcategories (see Table 23).

The first category focuses on the quantitative dimension of statements. In science the most desirable statements are universal statements, i.e. statements which always hold, regardless of context. Scientific theories and laws aim to be universal. In contrast, singular statements describe a single phenomenon. An example are facts which are needed to validate universal statements; in history singular statements are the main object of inquiry. While particular statements occur in scientific discourse, they are less important. They are easy to generate, low in information content and difficult to falsify. It is possible to automatically classify a statement into one of the three categories by examining the logical structure of a statement.

The second category differentiates between affirmative, negative and infinite statements. Statements of the first kind are more desirable than statements of the second kind; however it is not possible to ultimately confirm universal, affirmative statements. Statements of the third kind in this group (infinite statements) do not belong to the realm of science.

The third group concerns the type of relation described by the statement. Categorical relations can be further refined to include generalization relationships, part-of relationships and many other variants. Both hypothetical relations which include different variants of causal relationships and disjunctive statements are important for science.

Many problems of scientific knowledge are related to 'problematic' and 'apodeictic' statements which belong to the fourth category. 'Problematic' statements assert that something is possible while apodeictic statements are concerned with necessity. No completely satisfactory system of logic has yet been developed for both types of statements. While the natural sciences focus primarily on assertoric statements, in the social sciences, possibility and necessity also play an important role.

Table 23. Kinds of judgements

I. Quantity of Judgements	II. Quality
Universal [e.g. "All humans are mortal."]	**Affirmative** [e.g. "The body is mortal."]
Particular [e.g. "Some humans are Greek."]	**Negative** [e.g. "The soul is not mortal."]
Singular [e.g. "Socrates is Greek."]	**Infinite** [e.g. "The soul is in the unlimited sphere of immortal beings."]
III. Relation	IV. Modality
Categorical [e.g. relation of the predicate to the subject]	**Problematic** [e.g. "It is possible for animals to fly.]
Hypothetical [e.g. relation of the cause to its effect]	**Assertoric** [e.g. "Birds fly."]
Disjunctive [e.g. relation of subdivided knowledge and of the collected members of the subdivided knowledge together]	**Apodeictic** [e.g. "It is necessary that birds are animals."]

Source: Kant 1787, A:70, B:95.

These four categories determine the way in which a statement can be embedded in theories, the relationships it may have with other statements and the way in which the statement can be interpreted and evaluated. Therefore scientific knowledge management systems need to support the classification of statements into these four categories. When statements are described formally, this may be done automatically. An analysis of a statement in first order logic, for example, can often determine which type of relations are involved, which level of universality is present etc. For less formal statements, however, authors or reviewers may need to perform the categorization.

3.4.2.2. Subtypes of 'statement'

3.4.2.2.1. Axiom

In a collection of statements, axioms are those that cannot be derived from other statements in the same collection. It is possible that a collection can have several non-identical sets of axioms. When statements take the form of natural language instead of mathematical equations or logical propositions, then the term 'assumption' is often used instead of 'axiom'. Axioms may be formed inductively from observations and facts and therefore may have empirical basis. Alternatively, they may be deduced from collections of statements thought to be true or they may merely be useful assumptions without direct empirical basis.

3.4.2.2.2. Definition

Definitions allow to distinguish one concept from other concepts. They are useful to categorize given phenomena and to determine the corresponding linguistic concept. In their simplest form, definitions are identification rules which describe the boundary of a concept. Definitions only contain the minimum set of characteristics needed to distinguish instances of a concept from other phenomena. Definitions may also be based on sets of operations for identifying the concept rather than on sets of characteristics (Bridgeman 1991, p 58).

Definitions are relational in nature. One concept (the 'definiendum') is defined through a statement, the definiens. The latter must consist of already established concepts. The model of scientific knowledge, therefore, contains two types of relationships between concepts and statements: defined-by and occurs-in. The 'occurs-in' relationship identifies concepts which are part of the definiens. Linking concepts in a network of such relationships has the advantage that circular definitions can be recognized more easily and that central concepts within a domain can be identified. The latter may prove particularly useful when combined with the separation of homonyms. For example, when different interpretations of the term 'knowledge' are treated as different concepts then an analysis of definitions will quickly show which of these interpretations are more or less central.

Definitions vary in their completeness. Definitions are complete if the meaning of definiendum and definiens are identical. In such cases, the definition can be used as substitute for the concept. In mathematics, for example, definitions are often introduced as convenient shorthand notation. In science, however, concepts for which a complete definition exists are of secondary importance compared to original concepts which lack a complete definition. This is because they leave no questions open and because their semantics can be fully derived from other – and therefore more important – concepts. Of course, it is an important scientific accomplishment when concepts which were thought to be original can be completely derived from other concepts.

In addition, definitions differ in structure. Most definitions are circumferential. They describe the boundary of a concept by supplying some necessary and sufficient characteristics of an instance. An alternative are prototypical definitions. They describe core instances of a concept. The extent of difference between a phenomenon and the protytype(s) becomes the basis for determining whether the phenomenon fits the concept.

Various obstacles exist for achieving precise definitions: Because of the hermeneutical circle (see section 3.2.2.2. on page 91), it is not possible to achieve an unambiguous definition. Any definition is ultimately based on some concepts which are not clearly defined. Another problem is that the boundaries of original scientific concepts are susceptible to change. This phenomenon is not limited to the social sciences but also occurs in the natural sciences. Kuhn, for example, describes how core concepts such as time, mass, electricity etc. were subject to significant change even in essential properties (Kuhn 1996, p 102). Important concepts in management science such as business process reengineering (Hammer & Champy 1993) or group decision support systems have experienced similar change in meaning. While group decision support systems were initially regarded as a special type of decision support system, it soon became obvious that systems with such characteristics could be used for other kinds of tasks such as idea generation. An alternative term 'electronic meeting system' was introduced soon thereafter, but was only adopted at a slow pace by the research community. As a consequence, the term group decision support system first shifted from a special kind of decision support system targeted at groups to systems which support group work in a meeting context (including decision- and other tasks). Currently the term is undergoing a further shift towards electronic meeting systems focused at decision-making tasks. A third problem is that definitions of original concepts may be based upon assumptions about shared meaning in a community. Therefore definitions that may be appropriate for one subset of a community may not be adequate for another.

When evaluating a definition, the following rules need to be applied:

- Elimination of unnecessary elements. An element is unnecessary if the boundary of the concept does not change when the element is removed from the definition. Practical examples often occur in adjectives. Consider the following definition:

  "An entity is a (real or conceptual) object or an event." (Chen 1982 p 88)

  Objects are either real or conceptual, therefore both adjectives do not increase the boundary of the concept. The previous definition can be reduced to: "An entity is an object or an event."
- Avoiding circularity: Moore has pointed out that definitions may not include terms which require knowledge of the term which is being defined (Moore 1903, p 10). The term 'brother', therefore, cannot be defined as 'male sibling' (Scriven 1958, p 122). The term 'and' cannot be defined as "a particle expressing the general relation of connection or addition, used to conjoin word with word, phrase with phrase, clause with clause" (Scriven 1958, p 112). To understand the latter definition, an understanding of grammatical particles including 'and' is necessary.
- When definitions are not prototypical, only essential properties may be part of a definition. A property cannot be part of a definition, if it is meaningful to talk about an instance of the concept which does not have the property.

3.4.2.2.3. Proposition and hypothesis

A proposition is a statement that follows from a set of axioms (Babbie 1994, p 49). Hypotheses are specified expectations about empirical reality, derived from propositions (Babbie 1994, p 49). They are predictions about the values of empirical indicators which represent concepts that are part of proposition (Dubin 1978, p 206). Therefore the difference between hypotheses and propositions is only gradual. Hypotheses can be regarded as operationalized versions of propositions. For one proposition, it is usually possible to generate many hypotheses. To avoid unnecessary effort the scientific community needs to be efficient in testing hypotheses and avoid the testing of trivial hypotheses. One approach for accomplishing this goal is to standardize measures of concepts (Dubin 1978, pp 209-210). Bunge proposes the following criteria for the formulation of a proposition (Bunge 1998a, p 261; note that. Bunge uses the term 'factual hypothesis' instead of 'proposition'):

1. The proposition must be logically sound, i.e. formally correct and meaningful.
2. The proposition must be grounded on or compatible with previous knowledge (at least to some extent).
3. The proposition must be empirically testable (i.e. it must be possible to derive a testable hypothesis).

3.4.2.2.4. Facts

Some statements describe observations of phenomena from the 'real world'. Nagel defines facts as "singular statements that" ... "formulate the outcome of observations on the subject matter." (Nagel 1961, p 348) However, facts "are never simply given to us. They have to be accepted." (Miller 1997, p 70) Factual statements thus relate observed phenomena to the linguistic and abstract concepts on which science and language is based. Nagel emphasizes that the term does not imply commitment to a specific philosophical or psychological theory of observation. He also considers statements which describe procedures for data collection as facts. They are, however, better represented as relationships between a fact and a method.

A fact describes a single phenomenon. It "identifies one or more entities and attributes a property or relation to the entity or entities, with our without reference to time." (Little 1998) The name of a person, for example, would be considered a fact. Complex facts describe a combination of phenomena. A specific purchasing transaction which occurred at a specific time between a specific buyer and seller and involved specific objects, for example, is a complex fact. Sets of facts are usually referred to as 'data'. The description of complex facts requires several additional attributes. They are described in the section on data below.

3.4.2.3. Activities

Much of the complexity of scientific knowledge – if represented using the model of scientific knowledge discussed here – resides in statements. While concepts, theories, frameworks etc. provide entry points to the body of scientific knowledge, its detailed semantics arise primarily from the way in which the different components of knowledge are related to one another via statements. The variety of statement types and notations provide many opportunities for computer support and requires many custom solutions. Nevertheless some key services can be identified which are required for processing statements of all notation types.

3.4.2.3.1. Editing & Presentation

As the syntax of statements varies greatly between different types of statements it is necessary to provide a variety of software tools for the definition of statements. These tools should provide the capability to create, revise and delete statements and to perform simple syntax checks. In addition, they need to map statements to the other components of the meta structure, e.g. theories, concepts, problems etc.

Editors need to be provided for each of the four generic notation-based categories of statements. In each of these categories, several statement editors may be used which differ in their complexity as well in the subcategories of statements which are used. For example, procedural statements might be written in different programming languages, or in various forms of pseudo-code. Graphical statement editors might be based on free-form graphics – with little semantics – or implement a specific notation such as UML class diagrams.

Statement editors need to invoke other knowledge-related services. While a word processor could be used for creating an article in natural language, it is necessary to integrate the scientific concepts which are used in the text into the infrastructure. The integrating mechanism requires the identification of the core concepts of the article; in addition, it needs to search for overlap with existing concepts in the overall body of knowledge. If overlapping concepts exist, the user needs to map his concepts to the concepts in the infrastructure. Otherwise his concepts need to be added to the set of concepts within the knowledge infrastructure. Similarly, a knowledge infrastructure may incorporate a diagramming tool which supports object and process models. Such an editor needs to ensure that elements of the diagram are mapped to the corresponding elements in the model of scientific knowledge. The same applies to the graphical relationships; they need to be mapped to corresponding relationships in the model.

3.4.2.3.2. Evaluation

Different categories of statements need different approaches for consistency checking and other analysis. This may include theorem provers for logical statements, equation processors such as Mathematica and statistical software for mathematical statements, indexing and summarizing software for statements in natural language, etc. These processors may be an object of research themselves as they represent codified knowledge for evaluating knowledge.

3.4.2.3.3. Integration

When a new statement is created, it needs to be linked with several other elements of knowledge. Links need to be established, for example, to the concepts which occur within the statements. Ideally the link would be established at the creation of the equation, diagram or natural language statement. Integration, however, is complicated by the existence of concept variants. It may not be obvious which variant of concepts such as 'trust', 'user acceptance' or 'information system' the author of the equation has in mind. Therefore the author may need to spend significant additional effort in determining the correct concept variant to be linked to. In some cases, none of the existing variants may be adequate and an author may need to introduce a new variant. The significant effort that is needed for sorting out relationships between new statements and the existing body of knowledge has another consequence: authors may not fully integrate their statements into the existing knowledge network.

Therefore automated algorithms are needed which can identify potential interdependencies between a statement and the present network of knowledge. This analysis can be based on lexical characteristics of the terms used in the statement or on the attributes and the context of the statement. Lexical analysis extracts the natural language terms which occur in a statement (independently of whether this is a diagram, formula or natural language statement) and then searches for similar terms in the list of concepts. It then provides a list of candidate synonyms from which the author could then select the appropriate ones (Hars 1994, pp 180-193). Another alternative is an attribute-based comparison which searches for concepts which have similar properties or relationships.

Once links with existing concepts, theories, methods etc. have been identified, it is necessary to check the consistency of the new statement with already existing knowledge. Inconsistencies need to be flagged and – if possible – removed.

3.4.2.3.4. Generation

Statements can be used to generate other statements. Variables may be replaced with constants or constants may be replaced tentatively with variables. For example, economic models which deal only with a single product may be extended to include an indefinite number of products. They can then be extended to incorporate multiple periods etc. Thus one model can be used to automatically generate another model which can then automatically be checked for tractability. Graphical statements such as process diagrams could be transformed into a simulation model. Sophisticated generation approaches would allow the user to specify generation constraints. In addition, generation functionality can populate an abstract model with data and then perform an analysis.

3.4.3. Theory

The creation and validation of theories is a key aspect of scientific activity. Kuhn describes science as a battleground of competing theories in which only those with the highest explanatory power survive. A theory is a "linguistic device used to organize a complex empirical world" (Bacharach 1989, p 496) as well as to "explain and predict phenomena." (Emory 1985, p 30) Sutherland defines a theory as "an ordered set of assertions about a generic behavior or structure assumed to hold throughout a significantly broad range of specific instances" (Sutherland 1975, p 9). This definition emphasizes generality and broad scope as desirable characteristics of a theory. In addition, it identifies assertions (or, more generally, statements) as main components of a theory. The term theory, is similar to the term 'system' which usually is described as consisting of elements and relationships. It is also similar to the term 'model' (Dubin 1978, p 17-18). Therefore, for the purposes of this work, these three terms are treated as synonyms. For a rich range of controversial conceptualizations of theories see Agger (1989) who argues that "Theory in the best sense is speculative reason open to its own narrativeness – its fictive imagination" (p 117).

Theories need to be distinguished from empirical generalizations or laws. "Laws are universal generalizations whose non-logical vocabulary consists only of observation terms." (Strike & Posner 1983, p 349) An example for a law is the statement that "when water in an open container is heated it eventually evaporates" (Nagel 1961, p 79). The concepts which form part of this statement (water, boiling, temperature) are directly observable. In contrast, theories always contain some concepts which are not directly observable. An example is the concept of 'force' which plays a key role in Newtonian mechanics. The distinction between observational terms and theoretical terms is not sharp and therefore cannot lead to a sharp distinction between laws and theories. Nagel identifies a second difference: While theories usually consist of several statements, laws almost always consist of a single statement. In the model of scientific knowledge, laws therefore are represented by the 'statement' class while a separate 'theory'- class is needed to group the statements.

An example for a theory in information systems is the Technology Acceptance Model (TAM). It was first proposed by Davis (1989) who derived it from the Theory of Reasoned Action. It has been investigated widely in information systems research and several variants and extensions of the theory have been proposed (Gefen & Straub 1997). The theory consists of five core variables: Behavior (B), Behavioral Intention (BI), Acceptance (A), Perceived Ease of Use (PE), Perceived Usefulness (PU). Furthermore, it consists of merely three statements:

(1) $B = BI = w_1 \cdot A + w_2 \cdot PU$

(2) $A = w_3 \cdot PU + w_4 \cdot PE$

(3) $PU = w_5 \cdot PE$

The theory states that behavioral intention is a function of two variables, user acceptance and perceived usefulness. Perceived usefulness, in turn depends on perceived ease of use, which together with perceived ease of use determines the level of acceptance. Thus the core of the theory consists of three statements which in turn involve five different concepts.

3.4.3.1. Attributes and relationships

3.4.3.1.1. Overview

Several researchers have examined the components of theories. Dubin has developed one of the most detailed conceptualizations of 'theory' in the social sciences (Dubin 1978, see also section 3.3.3.3. on page 116). He distinguished seven components (see Table 24). The main components are 'units' and different types of statements. Units are 'properties of things'. They can be observational or theoretical terms and are equivalent to the term 'concept'. The relationships between units are reflected in statements which describe laws of interaction. Another type of statement describes internal and external boundaries of a theory. Boundary statements and interaction statements may overlap. The behavior of a system may vary depending on its overall state. An example is a market system which may exhibit the states of perfect competition, imperfect competition or monopoly. Therefore Dubin requires the identification of those persistent system states in which the laws of interaction are valid. Dubin's system states cannot be partial. Moreover, theories do not require the existence of states. A state can be regarded as a qualifier which is ascribed to a system if certain conditions hold. Therefore it can be subsumed under the 'concept' entity: A pricing theory may be modeled as being dependent on the concept of 'perfect competition' and thus differs from a pricing theory which requires a monopoly. An advantage of using 'concepts' to represent system states is the ability to represent partial states. Propositions and hypotheses are special kinds of statements which relate the units of the theory. Propositions follow from the laws of interaction and hypotheses are operationalizations of propositions. Empirical indicators are reflected in the meta structure as measures which are special types of concepts. Therefore all of Dubin's components can be mapped into the meta structure of scientific knowledge. Laws of interaction, boundaries, propositions and hypotheses are different types of statements. System states can be represented as a combination of state concept and associated statements. Empirical indicators correspond to measures and variables.

Table 24. Components of a theory

#	Component	Description	Meta structure equivalent
1	Unit	Concepts (real and nominal) which are at the core of the theory	Concept
2	Law of interaction	Statements which formally describe the interaction	Statements
3	Boundaries	Describes the (external) context in which the laws of interaction hold	Constraints (subtypes of statements)
4	System state	States within which the units interact differently	Represented as different theories with same elements but different constraints
5	Proposition	A truth statement (usually a subset or consequence of the laws of interaction)	Proposition (subtype of statement)
6	Empirical indicators	Empirical indicators related to terms in the propositions	Variables, measures (subtype of concept)
7	Testable hypothesis	Translation of the proposition into testable hypothesis	Hypothesis (subtype of statement)

Source: Dubin 1978.

Other researchers conceptualize theories as consisting of different components. Nagel(1961, p 90), for example, views theories as consisting of an abstract calculus, a mechanism for mapping observations to that calculus and an interpretation of the calculus. Walls, Widmeyer and El Sawy (1992) describe components of a special class of theories: design theories. They point out that in contrast to natural science, the goals of subjects must be taken into account in the social sciences. However, it is possible to include goals as additional expressions into the formulation of a theory or to describe goals as part of the context. Therefore an extension of the meta-structure to accommodate goals is not necessary. Hempel (1958, p.46) conceptualizes theories as consisting of "a set of sentences expressed in terms of a specific vocabulary" combined with "a set of interpretive sentences". He distinguishes primitive sentences (axioms, postulates) from derivative sentences or theorems. Similarly, the vocabulary is divided into "primitive terms - those for which no definition is specified - and defined terms." (Hempel 1958, p.46) These elements require the existence of a linguistic framework with a clearly specified logical structure to allow deductive inference.

Table 25. Attributes of 'theory'

Tag	Description
Title	Title
Description	Short description of the theory (summary)
Content	
↳ Background	Description of the context of the theory
↳ Detail	Detailed description of content, working and interpretation of the theory
↳ Limitations	Description of limitations which other researchers need to be aware of when applying this theory.
Type	
↳ Scientific goal	Prediction or understanding.
↳ Analytical focus	Interaction or outcomes
Quality	
Correctness	Level of confidence in the truth of the theory
Falsifiability	Whether theory lends itself to empirical falsification
Generality	Level of generality
Depth	Richness of detail
Parsimony	Simplicity of the theory
Utility	Usefulness of the theory
Fit	Fit of the theory with the existing body of scientific knowledge

3.4.3.1.2. Content

As Table 24 shows, the main parts of a theory are of a relational nature. Theories are groups of statements which together form a system and which contain concepts. Theories include other statements such as hypotheses and propositions and they may be linked to problems, facts etc. As a consequence, there are few requirements for extending the 'content' attribute of the 'theory' class. It is often desirable to describe the context in which a theory was developed. This can be done by a 'background' property. Another attribute is needed to capture detailed information about a theory in verbal format. This is useful for explaining the information that is formally represented in the relationships with statements, problems, methods etc. While information contained in the relationships is always accessible to the reader, an explicit discussion which brings the different elements together is useful for the reader. A third attribute concerns the known limitations of the theory.

3.4.3.1.3. Type: Scientific goal and analytical focus

Dubin describes additional characteristics of a theory. He argues that theories vary significantly depending on their goals: they can focus on prediction of a system state or they can focus on understanding the workings of a system. In the first case, the focus is on outcomes, in the second case, the focus is on the interactions and processes within the system. While one might expect that analysis of interaction advances the ability to predict or that an analysis of the outcomes might increase the understanding, he argues that this is rare in scientific practice. As shown in Table 26, of the four possible combinations, only two are predominant.

Table 26. Relationship between goals of research and analytical focus

Analytical focus	Goals: Understanding	Goals: Prediction
Interaction	x	
Outcomes		x

Source: Dubin 1978, p 20.

Theories which treat a system as a black box have the advantage of describing patterns of relationships between observable phenomena. Although it is not possible to exclude that the patterns have been caused by other factors, careful research design can significantly reduce this problem. However, such theories strongly depend on the selection of the set of phenomena which has been observed. Whether they can be applied to a specific case depends on whether the set of phenomena is representative for the case. While it can be shown that a theory is representative of the collection of all known applicable cases, it is not possible ex ante to decide whether a subset is representative of the collection of all known cases. Patterns which are valid for a typical enterprise, for example, may not apply in specific industries. Unless a theory has been controlled for 'industry', then, the application of findings of the theory to a company of an industry may be fraught with problems. A similar problem does not occur for a theory which is focused on interaction. As the theory describes how different elements of a system cooperate the applicability of the theory can be identified by examining the characteristics of the elements and their relationships. Interaction theories, however, have the drawback that other interactions may occur in parallel which may affect the outcome. This reduces the ability to empirically test the accuracy of the theory and may lead to cases in which incompatible theories for the same phenomenon are advanced without being able to decide which is more accurate.

An important difference between both types of theories, however originates in the process of theory building and evaluation. To build a predictive theory, sets of phenomena must be identified which are suitable for statistical analysis and care must be taken to ensure that the phenomena are generalizable and that no obvious factors exist which might lead to systematic deviations in the patterns among its subsets. The results of the theory must be interpreted carefully to avoid such mistakes. The development of a predictive theory, therefore is strongly based on statistical analysis.

On the other hand, theories which focus on understanding must proceed differently. They must strive to identify a mechanism which produces certain behavior. A major element of such research is to identify the details of the mechanism and to conclusively demonstrate the working of the (abstract) mechanism. A second and clearly separable step would be to search for evidence that the mechanism is active in the hypothesized settings. Once the mechanism has been identified and shown to be active in certain conditions, an analysis of its predictive value needs to follow. As a consequence, interaction theories have the disadvantage of a much longer lead time until they can find support from empirical observation. This is an additional reason why interaction theories are less in demand than predictive theories.

Because of the different research approaches and evaluation procedures which are associated with the focus of theories, the distinction needs to be incorporated in a knowledge infrastructure. As part of the specification of a theory, the author should be required to define analytical focus and goals. For the design of the system it is also necessary to determine whether the other two combinations (focusing on interaction with the goal of prediction or focusing on outcome with the goal of understanding) can be discarded as irrelevant. While we agree with the assessment of current practice, we cannot exclude theories which advance both the understanding of interaction and outcome. Therefore both types cannot be merged into one.

3.4.3.1.4. Quality

3.4.3.1.4.1. Correctness

Ideally theories should have a property which allows to distinguish between true and false theories. Unfortunately, as Popper has pointed out, the truth of a theory can never be proven. A number of obstacles stand in the way of absolute truth: A first problem is the use of inductive reasoning: Proof of a theory would either require the examination of all cases in which the theory applies to ensure that no counterexample exists or the deduction from another proven theory. The former is not possible for theories which claim relevance for the future. The latter is not possible as no absolutely proven theories exist. Another problem arises from the hermeneutical circle. The concepts on which theories are based lack a solid terminological foundation. They can never be defined with full precision. Therefore the meaning of a theory can never be stated unequivocally which would be a prerequisite for being able to assign an unequivocal truth value (For a detailed discussion of these issues see section 3.2.2.1. which begins on page 91). Because of these problems, most scientists now agree that ultimate certainty about the truth of a theory can never be reached if the theory incorporates real-world phenomena.

Another problem is deciding when a theory is to be regarded as false. Ideally, theories should be discarded when contradicting phenomena are found. This principle has been challenged on two grounds: Strict application of the principle may lead to the rejection of almost all known theories because theories without exceptions are indeed the exception. Kuhn has pointed out that theories often contain known problems which scientists hope to eliminate over time (Kuhn 1986). In the practice of science, therefore, theories are not discarded once inconsistencies with observation arise. Of course, such cases weaken a theory and the more problems are identified, the more pressing the need becomes to find alternative theories. A second objection against discarding such theories is the ability of scientists to correct a theory to account for mismatches with observation. Such a practice is widely used in respected fields such as quantum physics. As a consequence, exceptions and counterexamples are criteria which can be used to compare theories rather than to accept or reject individual theories.

Therefore, it is not possible to add a binary 'truth' attribute to the 'theory' class. An alternative is the introduction of a continuous truth variable that allows values between 0 and 1, where 0 represents a false and 1 a true theory. Values between 0 and 1 represent different degrees of confidence in the correctness of a theory. In the following the term 'correctness' will be used instead of truth to avoid the connotation of absolute right or wrong that readers often associate with the latter. It must also be possible that theories have no truth value (i.e. when they have not yet been tested or evaluated by others).

Several approaches can be applied to estimate the correctness of a theory. One approach is to ask reviewers to evaluate the theory and to aggregate their assessment into a single correctness value. This approach has the drawback that theories which are based on a new paradigm may be rated as incorrect as long as the scientific community has not shifted towards the new paradigm. A more indirect way of assessing the correctness of a theory would be examining the network of other concepts, theories and data in which the theory is embedded. The confidence in a theory would increase with a larger number of tests which the theory has passed and a larger number of other theories in which it is embedded (and which also have passed empirical tests). While it is not possible to confirm a theory, a large number of different cases in which the theory is successful increases the confidence in a theory. Confidence in a theory grows with the number of persons who have applied it, with its age and with the number of studies which have examined it without finding major problems. On the other hand, the confidence in a theory can be quickly withdrawn when observations contradict the theory or when a better theory emerges. A knowledge infrastructure allows to automatically determine some of the factors on which the confidence level for a theory depends. This includes the number of studies which have investigated the theory, their results, the number of persons which apply the theory and the number of other theories which successfully rely on this theory.

3.4.3.1.4.2. Falsifiability

Because of the fundamental problems associated with absolute truth, Popper proposed falsifiability as a criterion for assessing scientific theories. He emphasized that "it must be possible for an empirical scientific system to be refuted by experience." (Popper 1968, p 41) Organization theory abounds with examples of theories which are stated so vaguely that they are almost impossible to falsify (Bacharach 1989, p 501). Falsifiability requires clearly defined, operational concepts, precise statements which relate the concepts and hypotheses which are rooted in fact rather than definition. In an electronic environment these problems are easier to detect than in a paper environment: It is possible to automatically compute an index which describes the number of new concepts defined for the theory compared to the number of concepts which have been reused. The larger the number of concepts which are part of other theories, the higher the chance of falsification and the higher may be the value of the theory. Similarly, it is possible to compute the number of operational concepts. It is more difficult to automatically assess the precision of statements with the exception of distinguishing between mathematical and logical statements on one extreme and graphical, linguistic statements on the other. Hypotheses can be analyzed to identify to what degree they focus on operational factors or on definitions. While the falsifiability of a theory is a criterion for rejecting some theories which are overly vague, Kuhn has pointed out that the criterion suffers from similar deficiencies as 'absolute truth':

> "If any and every failure to fit were ground for theory rejection, all theories ought to be rejected at all times. On the other hand, if only severe failure to fit justifies theory rejection, then the Popperian will require some criterion of 'improbability' or of 'degree of falsification'" (Kuhn 1996, pp 146f)

3.4.3.1.4.3. Generality

Theories differ in the range of phenomena which can be covered. Theories are preferable if they are applicable to a broad number of phenomena (Bacharach 1989, p 509); this also distinguishes theories from experimental laws (Nagel 1961, p 89). More general theories achieve a higher unification of intellectual concepts. They are more 'fundamental' (Davies 1965, p 40). Often a tradeoff exists between the generality of a theory and the specificity of insights which can be derived from it for each individual phenomenon. Boulding eloquently described the problem as follows:

> "We always pay for generality by sacrificing content and all we can say about practically everything is almost nothing." (Boulding 1956)

This is especially relevant for the social sciences where theories which are comparable to Newton's mechanics or Einstein's relativity are hard to find. As a consequence, researchers in the social sciences have advocated the development of 'middle range theories' which are "solutions to problems that contain a limited number of assumptions and considerable accuracy and detail in the problem specification." (Weick 1989, p 521) Measurement of 'range' is difficult. It requires a conceptual system which orders concepts by level of generality. The model of scientific knowledge shows a hierarchical relationship which can be used to compute different levels of generality.

3.4.3.1.4.4. Depth

Besides the range of phenomena covered, theories need to be differentiated by the richness of detail which they provide about a phenomenon. A theory has greater depth if the amount of information which it can generate for a phenomenon is larger. For example, if a phenomenon is associated with a fixed number of variables, then the theory with the larger subset of dependent variables is deeper. A similar argument holds for independent variables. A theory has more depth if it can take a larger number of independent variables with the same set of dependent variables into account. Unfortunately, the frames of reference often change between theories; the sets of attributes used to characterize a phenomenon may change. In many cases, a new theory can derive attributes of an older theory and therefore should be categorized as deeper. In a knowledge infrastructure where previous conceptualizations of a phenomenon are not discarded (but unnecessary attributes can be determined automatically) this increase in depth can be identified with ease. A second aspect of depth is

> "the precision of the predictions that can be made from it [a theory]. If the theory predicts that some experimental variable x equals A under certain conditions, it is highly precise (though maybe not right). In descending order of precision, for example, are predictions: x = A; B > x > A; x > A; x is significant; some x may occur in the next few days." (Davies 1965, p 40)

3.4.3.1.4.5. Parsimony

Theories should be simple. Among two theories with the same set of conclusions, the less complex is preferable. An extension of this principle is Ockham's razor which is based on the distinction between important and non-essential conclusions. Two theories should be compared on the basis of their important conclusions only. Ockham's razor thus leads to the simplification of theories by eliminating inessential insights. However, the distinction between both types of conclusions is often not obvious. Another aspect of parsimony is a preference for simplifying theories to expose their core ideas. In many cases the mechanisms for extending a theory to cover additional phenomena or to increase its detail are obvious. An investment theory which covers multiple periods and a single product can often be extended to cover several products or to include probabilities without changing the basic approach. In cases where possible extensions are obvious, it is preferable to expose the core of the theory rather than to provide the most universal theory.

Bacharach proposes the ratio of hypotheses to propositions as a measure for parsimony. "A theory where each proposition covers five hypotheses is preferable to one where each proposition only covers two hypotheses" (Bacharach 1989, p 508). However, this approach has significant shortcomings: in practice many propositions which follow from a theory remain unstated. Similarly, the effort which a research team can spend on the evaluation of a theory may dictate the number of explicitly formulated hypotheses. Thus the ratio of explicit propositions and hypotheses may be misleading. If the analysis were to be extended to all propositions and hypotheses that can be formulated then several problems would arise: When are two hypotheses or propositions to be counted as different? How are negative hypotheses and propositions to be treated? Is it necessary to exclude hypotheses which estimate specific values of variables because this would enable the generation of an infinite number of hypotheses for any proposition which functionally relates an independent and a dependent variable? Therefore Bacharach's measure is not suitable for measuring parsimony.

3.4.3.1.4.6. Utility

Lewin's statement that "nothing is quite so practical as a good theory" (Lewin 1945), indicates that theories should not just be evaluated on their accuracy or generality but also on the impact which they have on practice. As science progresses, the number of open questions multiplies; each new finding gives rise to many additional problems. Each researcher, therefore, is faced with a mountain of potential research issues and must choose on which to focus his efforts. Unfortunately, researchers are not always effective in this process. They are often criticized of spending their efforts on issues that have little practical value. In a situation where the number of theories as sets of true statements about our world is probably inifinite, the utility of a theory becomes an important characteristic because it ensures that researchers focus their attention on topics which are important (which does not imply agreement on importance). Although every researcher must make the choice and therefore – consciously or unconsciously – considers the value of different research alternatives, the systematic consideration of utility has been widely neglected by the research community. This is a consequence of the predominant positivist perspective which regards research as value-free.

Several criteria have been proposed to assess utility: the value which a theory has for practical application – even if those applications may be in the future. Theories should be original in that they should identify relationships that were not known or under serious doubt before. They should solve a problem, allow new practices or point out or open new fields to study. Mazur emphasizes this point when he declares that "the essential characteristic of science…is that it is profound – it is a body of theoretical knowledge that is not trivial." (Mazur 1971, p15) At minimum, they need to carry an element of surprise.

3.4.3.1.4.7. Fit

Another important aspect of a theory is its relationship to the body of preexisting theories (Bacharach 1989, p 511). A theory may replace existing theories by integrating their explanations into a consistent whole. This may involve showing that some of the conclusions of the earlier theories were incorrect. In other cases, a theory may extend other theories or it may 'bridge the gap between two or more different theories thus explaining something between the domains of previous theories" (Bacharach 1989, p 511). When a theory causes existing theories to be "reevaluated in a new light" (Bacharach 1989, p 511), then the theory is said to be transformational. Transformation frequently involves changed interpretations of concepts associated with the theories.

In an online infrastructure, the previous interpretations need not be discarded. Instead, a new version of the concept can be introduced which reflects the changed interpretation. When an old theory is reevaluated, this implies that an attempt is made to adapt an existing theory to the new concepts. This in turn, yields a new version of the older theory. Even if it can replace the previous version, the previous version need not be discarded but merely marked as superseded. The introduction of new theories, therefore, may significantly extend the network of existing concepts and theories by introducing many new versions of concepts and theories.

3.4.3.1.5. Relationships

Theories are systems which describe relationships between a set of concepts. These relationships are expressed in the form or mathematical, logical, graphical, procedural or linguistic statements. Figure 18 on page 121 shows an explicit relationship between theories and statements. The relationship between theories and concepts is implicit: As statements incorporate or define concepts, it is possible to extract a list of related concepts from a database which implements the conceptual model. A query can determine which concepts are involved in each statement. A knowledge infrastructure may provide different levels of detail for such queries. The result may vary depending on in the information needs of the user. It may include only the most central concepts which occur in the theory's statements. It may include all concepts which occur in those statements or be extended to include all concepts which are needed to define the concepts used in the theory etc.

Theories are related to most other elements of scientific knowledge. Dubin identifies additional relationships to propositions and hypotheses – which are subtypes of statements as well as to empirical indicators which are subtypes of concepts (Dubin 1978). Empirical indicators are represented by the 'measures' subclass of concepts. Furthermore, theories are related to observations – data and facts – via statements.

3.4.3.2. Activities

3.4.3.2.1. Editing & presentation

One of the opportunities for knowledge infrastructures are novel approaches for visualizing theories. In the paper-based environment, authors choose how to present their models. They may use equations, corresponding causal diagrams or other types of diagrams, etc. With information technology, equations can be automatically transformed into diagrams. Moreover, presentation techniques may allow to vary the level of detail in a diagram or a set of equations by hiding those aspects which are of little relevance. Diagram editors may be used to formulate theories. They may highlight dependencies on variables which already exist in the knowledge infrastructure and thereby promote the reuse of knowledge. Furthermore, presentation functionality is a key issue for the integrating mechanisms discussed below.

3.4.3.2.2. Integration

Theories consist of statements and concepts. They are related to problems, methods and facts. For the integration of theories, several additional issues arise. Interdependencies with other theories can be identified by analyzing similarities with sets of statements used by other theories and with the concepts which are relevant for already established theories. Once similar theories are identified, the relationship to the new theory can be determined. For example, the new theory may replace some older theories. In that case, the old theories can be marked as retired but need not be withdrawn from the infrastructure. Similarly, appropriate relationships between new and existing theories need to be added if the new theory is an extension or refinement of older theories. Sometimes theories are inconsistent. Some inconsistencies - although not all - can be recognized by analyzing the sets of statements belonging to the theories.

An additional function of integration is to alert members of the scientific community about changes in the status of theories. Researchers should be notified when a novel theory is introduced in one of their areas of expertise. Similarly when a theory is retired or extended, the authors of the theory and other researchers who are working on the topic need to be alerted.

Usually many teams of researchers work on aspects of the same problem domain or theories. As they formulate their results, they focus on their specific slice of the problem. Thus over time, many pieces of knowledge are created but few researchers devote their time to assembling these pieces into a consistent whole. Scientific journals are reluctant to publish such work because it is not considered 'original' research. This situation has significant disadvantages. It may take a long time until the 'big picture' for organizing the pieces within a problem area may appear and this may prevent individual teams from asking the right questions. Dependencies between different aspects of a problem may not be recognized and therefore teams may use incomplete problem formulations.

An online environment enables researchers to integrate their models and to monitor the effect of advances from other research teams on their own slice of a problem. Ideally the online environment would lead to an aggregation of individual research work into a more consistent whole.

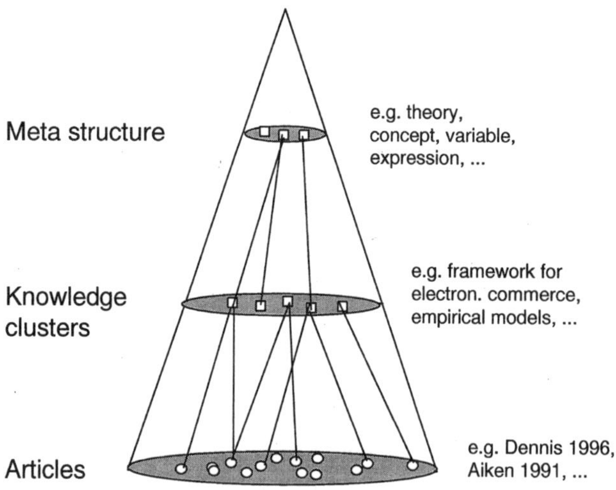

Figure 24. A hierarchy of scientific knowledge

Figure 24 illustrates this goal. A knowledge infrastructure can be thought to consist of different levels of scientific knowledge. On the bottom level, the level of articles, the contribution of each author is stored individually. Certainly, links such as references may exist between different articles which can be used to identify relationships and clusters of knowledge. However, it remains very difficult to identify the current body of knowledge for a topic from those contributions that have been discarded or that were early stepping stones. The emergence of a reliable body of knowledge requires the consolidation of knowledge around some common and recognizable elements. This is provided by the second and third levels in Figure 24. The second level contains frameworks, theories and models which are used by several authors and in several research efforts. Every contribution from the article level should be linked to corresponding units on the intermediate level.

In many cases such integration can occur automatically. When an author formulates an empirical model and describes its variables, the knowledge infrastructure can identify more generic models and theories in which these variables occur. In addition, the infrastructure may ask the author to manually establish the relationship to existing clusters of scientific knowledge. When no corresponding clusters exist, the researcher may be asked to create a new cluster to which the contributions of other researchers can be added thereafter. Such integration would increase the quality of contributions, as authors would need to clarify their differences from and overlap with current research more clearly.

An intermediate level of knowledge has significant advantages. A reader searching for knowledge concerning a particular topic would not be restricted to the article level where keyword searches and reference browsing are the primary means of analysis. He could search for theories or frameworks which deal with his topic. Subsequently he could navigate to the corresponding articles to pursue specific issues more in depth and to analyze competing and/or inconsistent findings. A researcher who embarks on a new project would be able to get direct access to all previous research which is related to the topic. The third level represented in Figure 24 contains the model of scientific knowledge as a meta structure.

An example shall illustrate the advantages of integrating knowledge. In information systems, electronic meeting systems (EMS) has been an active research area for more than a decade. Many empirical studies have been conducted to identify the success factors for EMS. For practical reasons, most studies have examined only a small set of variables out of the large set of variables which have been found relevant. Already in 1991, an overall theory for EMS (Aiken, Sheng & Vogel 1991) had been proposed but few of the subsequent empirical studies explicitly reference it. Empirical studies are rarely placed in the context of the whole theoretical framework. This lack of awareness has several drawbacks. Researchers who are empirically testing dependencies within the theory may not be aware of all of the other research within the same theory. They may develop instruments and approaches which some of their colleagues already have developed, thus unnecessarily duplicating work. In addition, testing of relationships within the theoretical framework is ad-hoc and fragmentary.

These problems could be avoided if a scientific knowledge infrastructure provided a consolidated theory for electronic meeting systems and if each article on EMS were linked into the theory at the time of being published. An example for such a theory is shown in Figure 25. It was derived post facto from an analysis of

16 EMS-related articles. The initial version of the theory was based on Aiken's et al's 1991 framework. Subsequently the models underlying the other studies were examined. Variables were mapped into the theory where possible, adjusting the terminology where needed. Some studies introduced additional variables which then lead to an extension of the original model. The resulting structure clearly shows which aspects of electronic meeting systems have been analyzed. It provides access to each original study and therefore would reduce the effort for further research projects. Moreover, the consolidated theory invites a complete analysis of the factors affecting EMS. It shows deficits; for example, very few studies have examined the importance of specific technical features for electronic meeting systems. It is to be expected that such support for integration will lead to the formulation of deeper theories.

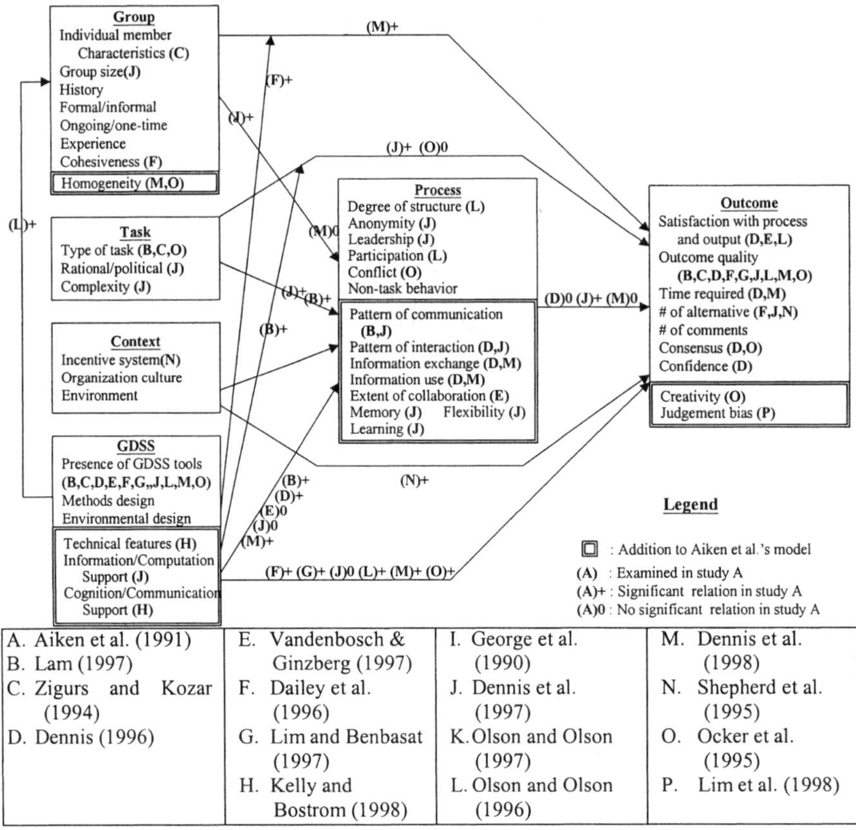

A. Aiken et al. (1991) B. Lam (1997) C. Zigurs and Kozar (1994) D. Dennis (1996)	E. Vandenbosch & Ginzberg (1997) F. Dailey et al. (1996) G. Lim and Benbasat (1997) H. Kelly and Bostrom (1998)	I. George et al. (1990) J. Dennis et al. (1997) K. Olson and Olson (1997) L. Olson and Olson (1996)	M. Dennis et al. (1998) N. Shepherd et al. (1995) O. Ocker et al. (1995) P. Lim et al. (1998)

Figure 25. Cumulative research framework for EMS
Source: Hars & Im 1998.

3.4.3.2.3. Evaluation

A large part of scientific research focuses on the evaluation of theories. While many different and domain-specific methods exist, most methods focus on a combination or refinement of the quality criteria which have been discussed in the section on quality attributes above. The result of the evaluations can either be recorded as static attribute of a theory or can be dynamically generated by analyzing the network of relationships in which a theory is embedded.

3.4.4. Problem

> "The term 'problem' designates a difficulty that cannot be solved automatically but requires a search, whether conceptual or empirical. A problem, then is the first in a link of a chain: Problem – Search – Solution." (Bunge 1998a, pp 193f)

Problems are the "spring of science."..."To stop handling problems is to stop research." (Bunge 1998a, p 187) Scientific progress is often equated with the solution of problems: Thus Kuhn argues that "the unit of scientific achievement is the solved problem." (Kuhn 1996, p 169) In the theories of knowledge, problems have been widely analyzed from a logical perspective: any hypothesis, statement or proposition with unknown truth value qualifies as a problem. As every new scientific finding or theory allows the creation of many new hypotheses, scientific progress leads to an explosion in the number of problems; for all practical purposes, the number of scientific problems is infinite. On the logical level, requirements for well-formedness have been described by Bunge (1998a, p 197-205). However, many scientific problems only reach this state once they are solved; therefore most active problems do not meet these criteria.

In scientific practice, the development of worthy problems frequently consumes a significant share of the overall research effort. "It is not unusual for a researcher to give six months to a year to finding a specific research problem and formulating it for research study." (Miller 1991, p 17) The clear formulation of a problem often constitutes the main breakthrough in a research project (Franklin & Osborn1971, p 93). However, the epistemological status of 'problem' is not well established. In epistemology as well as in scientific practice the processes of problem identification, maturation and selection have not received enough attention (Laudan 1996, p 79; Alford 1998, p 25; Greer 1971, p 96). Problem formulation is a creative process which is difficult to predict or manage and therefore is all too frequently excluded from scientific analyses and regarded in some way as a 'pre-scientific' activity.

This has significant drawbacks. Weick, for example, regards "fuller problem statements" (Weick 1989, p 517) as essential for the development of better theories. After reviewing shortcomings of current research practice, Dunnette recommends:

> "..a more careful and studied choice of research questions."... "Research energy should be directed towards questions that contain as few as possible of any prior unproven assumptions."..."We must be constantly alert to the narrowing of the research perspectives due to prior theoretical or methodological commitments." (Dunnette 1978, p 276)

Currently no adequate mechanisms have been established in the scientific community for problem identification and prioritization. The primary communication channels – scientific journals – only evaluate problems indirectly once they are solved. They don't explicitly support discussions on research problems. Similarly, funding agencies have developed ad-hoc approaches but no universal mechanism has emerged for the identification and formulation of problems which applies to the scientific community as a whole. This situation is sub-optimal and better mechanisms are needed for allocating research capacity to scientific problems.

The consequence for a practical theory of knowledge is that problems need to be incorporated in their various stages of maturity. In addition, a knowledge infrastructure needs to provide mechanisms for the collaborative prioritization of problems. This requires, for example, that members of the community are able to define, discuss and prioritize problems.

Some disciplines, for example Operations Research, have been successful in channeling research efforts by identifying core problems (e.g. the "Traveling Salesman Problem"). It would be useful to apply this approach to other disciplines as well: What, for example, are the key problems in information systems?

3.4.4.1. Attributes and relationships

3.4.4.1.1. Overview

Table 27 lists attributes needed to describe a problem. Besides attributes which are common to all knowledge contributions such as 'title' and 'description' only few additional attributes are needed to characterize problems. Similarly to concepts, an attribute is needed which distinguishes the level of abstraction. It allows the distinction between observational problems which lend themselves to an empirical approach or theoretical problems which require a conceptual approach.

A 'status'-attribute is needed to describe whether the problem is still open or whether it has been solved. Another attribute is needed to distinguish between degrees of specificity. Initial problem statements can be very vague and research aims to replace them with more precise, well-formed problem formulations. While a stock of fuzzy, poorly formulated problems is always available at the frontiers of science, a methodical search for a solution can only begin once a problem has achieved a certain degree of specificity. However, research activity – and with it the focus of a knowledge management infrastructure - cannot be restricted to the latter part. Sometimes significant advances in science do not occur from pursuing well-described (but unsolved problems) but rather from formulating novel but little understood problems. An important derived attribute for problems are solution approaches. They can be sketches of a method, theory or study which can be indirectly related via the statement class.

3.4.4.1.2. Quality

3.4.4.1.2.1. Relevance

A key criterion for a problem is its relevance. When a problem is added, the author needs to argue about the impact of its solution. A problem must identify a phenomenon that is in doubt and that cannot be explained with current theory

(Schauenberg 1998, p 52). It must be focused enough to be separable from other problems. A problem has low relevance if there is little disagreement about its solution in practice and if at the same time the researcher has little reason to expect that his results will change the established conception. A problem statement should contain a description of the potential impact of a solution.

Table 27. Attributes of 'problem'

Tag	Description
Title	Title
Description	Abstract
Content	
↳Detail	Detailed description of the problem.
↳/Solution approaches	Approaches for the solution of the problem.
Subject	
↳Level of abstraction	Theoretical/nominal vs. observational/real term
Type	
↳Degree of specificity	How clearly defined the problem description is.
Quality	
Relevance	Potential impact the solution of the problem may have
Clarity	Clarity of the description of a solution space
Solvability	Extent to which problem appears solvable with current scientific tools
Priority	How important is this problem compared to other scientific problems?
Status	Degree to which the problem has been solved

3.4.4.1.2.2. Clarity

A problem statement must clearly distinguish current knowledge about the phenomenon from the aspects which are under doubt. While a problem statement need not sketch potential solutions, it must identify the solution space with sufficient precision as to be able to subsequently recognize a solution. As a consequence, a problem statement must always outline some properties of a solution. In addition, a problem statement should contain a description of the potential impact of a solution. While the impact always contains a subjective element, it is an indispensable basis for the prioritization of research efforts.

3.4.4.1.2.3. Solvability

Although problems need to be relevant and should represent an important need, Kuhn has pointed out that in normal science the most important characteristic of a problem may be that it is solvable with the existing scientific tools. He cautions:

> "An excessive concern with real problems, regardless of their relation to existing knowledge and technique, can so easily inhibit scientific development." (Kuhn 1996, p 96)

Therefore a balance needs to be found between relevance and solvability.

3.4.4.1.3. Relationships

Problems are always related to other elements of scientific knowledge such as concepts, theories, methods, data or even other problems Bunge (1998a, p 194) calls this the 'background' of a problem. They need not always be related to a theory or a method because problems are often motivated by the absence of a theory to explain a certain phenomenon. In the model of scientific knowledge, the link between problems and other elements of knowledge is represented via the 'raises' relationship between 'idea' and 'problem'.

In scientific practice, problem descriptions are usually combined with an enumeration of potential approaches for their solution. Approaches are special cases of methods. Therefore solution approaches need to be modeled as relationships to corresponding methods rather than as an attribute of a problem. Thus problems are intricately intertwined with methods.

Problems are often broken down into more detailed problems. This can be described via the standard contains and is-a relationships.

3.4.4.2. Activities

3.4.4.2.1. Editing and presentation

Problems provide few additional requirements for editing and presentation. They can be described in natural language, much as theories and methods. They may consist of statements which requires a link to the corresponding statement editors. When problems originate in a specific part of a theory, method, data, etc. the author should be able to graphically point towards the problematic areas; the editor should then generate the corresponding links automatically. Conversely, when a reader accesses a problematic element of knowledge, the presentation system should be able to highlight the problem areas and display the detailed problem description on demand.

Another important functionality is the ability to sort problems flexibly by priority, subject area, type etc. The editor should allow to decompose a problem into various sub-problems and the presentation system should be able to indicate graphically which of the sub-problems are still open and which have already been completed.

A third aspect of editing and presentation concerns the automatic generation of problems. Heuristics may be developed which analyze the content of the knowledge infrastructure and which create problem entries when they encounter certain types of inconsistencies.

3.4.4.2.2. Integration

A major aspect of defining – and solving a problem – is to determine its relationship with other elements of knowledge. Therefore functions for identifying frameworks, theories, methods, concepts, etc. that are related or that may offer a solution is a core service of knowledge infrastructures. Because of the initially imprecise nature of problems such functions need to be interactive. Approaches on which such functionality may be based can range from analyzing the network to determine a core set of directly relevant concepts and theories to associative searches which aim to detect solved problems from other areas with some similarities in patterns. Such integration functionality may benefit from collaborative filtering approaches which allow to distinguish the most useful associative heuristics.

A second requirement for integration is functionality which monitors the infrastructure to alert researchers when critical subproblems have been solved or when advances or modifications in theories and concepts occur that may have an impact on the problem.

3.4.5. Framework

In the course of research, scientists frequently need to establish conventions and perspectives from which to analyze phenomena. Frameworks are tools which organize scientific concepts. In contrast to theories, the semantics of their elements and their relationships are rarely clearly established. Rather than explaining a phenomenon, frameworks propose a useful way for thinking about a problem domain. They are often the precursor or foundation of theories. A theory may explain the phenomena in a domain by developing a model on the basis of one the frameworks which have been proposed for that domain. The theory then uses the concepts and boundaries which have been established in the framework.

A framework is an organized collection of concepts. The emphasis is on the boundaries between concepts and on basic interdependencies. An example are frameworks for electronic commerce which distinguish between business-to-business, customer-to-business and intra-organizational electronic commerce (Kalakota & Whinston 1996, p 218). The value of such a framework lies not so much in the theoretical claim that the three concepts differ but in its usefulness for drawing boundaries and selecting research problems in the large conceptual space of electronic commerce. Thus many researchers have used this framework to position their research projects.

The boundaries between frameworks and other elements of knowledge are fluent. Taxonomy and typology are similar in intent and therefore will be subsumed under the framework category. As frameworks are extended and become increasingly detailed, they may gradually be transformed into a theory or a method. In the model of scientific knowledge this would be achieved by an increasing number of statements which constrain the framework and which define its interdependencies with other elements of knowledge. Thus the transition to theory can in part be recognized via automated algorithms.

3.4.5.1. Attributes and relationships

3.4.5.1.1. Title

Similarly to concepts, frameworks need a distinction between a unique long name and a more convenient short name (alias) which can be used when the context is clear.

3.4.5.1.2. Content

The content attribute captures the description of the elements and of their specific relationships. Frameworks often exhibit similar structure. Many frameworks, for example, take the form of a matrix with two or more presumably independent dimensions. The cells contain concepts which are relevant for each combination. The simplest structure is a list of concepts which are relevant for the specific context. Such frameworks do not contain assumptions about specific relationships between elements. They can be critiqued mainly on hidden overlap between some of the elements as well as on incompleteness. The structure of a framework can be described in the meta structure attribute.

Table 28. Attributes of 'framework'

Tag	Description
Title	
↳Long name	Name which uniquely identifies the framework
↳Alias	Alternative term for referring to the framework (may overlap with other terms)
Content	Content of the framework corresponding to the predefined structure
↳Meta structure	Type of graphical or mathematical layout
Quality	
↳Originality	Extent to which the framework differs from other elements in the infrastructure
↳Completeness	Extent to which all relevant phenomena are covered
↳Significance	Usefulness of the framework

3.4.5.1.3. Quality

Similarly to the 'concept' class, originality and significance are key criteria for the quality of frameworks. Originality requires the absence of other frameworks which cover the same domain. Significance needs relationships between the framework and existing concepts and theories. In the long run, the significance of a framework is lower, if the framework relies on fewer established concepts. Its significance increases as the number of theories, methods and studies which incorporate the framework increase. Significance may be computed automatically by analyzing the network of concepts. However, the measure may fail for assessing novel, highly original frameworks for which corresponding theories and concepts are only being established. An additional quality criterion is completeness. Frameworks should reflect all relevant phenomena within the framework's perspective. In some cases completeness-related problems can be detected automatically by analyzing the network of concepts. For example, when a framework contains most but not all subclasses of a more general concept, this may be an indication of a lack of completeness. Another indication of incompleteness may occur when a framework contains mostly pairs of concepts and their opposites but fails to include some opposites.

3.4.5.1.4. Relationships

Frameworks are always linked to concepts which occur within the framework. These relationships are established indirectly via statements. Frameworks are more valuable if they rely on established concepts which are not introduced by the framework. In an electronic environment, it is possible to computationally determine the extent to which a framework relies on established concepts (e.g. defined or used by persons other than the framework's authors) and to which extent the framework defines new concepts as part of the framework. Frameworks need not be related to theories. Often frameworks will be created during problem specification.

3.4.5.2. Activities

3.4.5.2.1. Editing and presentation

As explained above, frameworks often adopt standard meta structures. Therefore editors are needed which are able to create and present frameworks in matrix-, architecture or other formats. These editors need to include not only the diagramming component but also need to have functionality for ensuring that the framework does not violate the meta structure semantics. Frameworks often change rapidly as the knowledge about the underlying domain increases. Therefore mechanisms for updating frameworks, appending elements as well as version management are very important.

3.4.5.2.2. Evaluation

Frameworks provide guidance within a specific domain. They take a specific, often subjective perspective on a set of issues. Therefore truthfulness of a framework is not a primary criterion. Originality, completeness and significance are more important. These attributes can be identified through a combination of interactive and automated approaches. Interactive approaches allow readers and reviewers to assess the framework; automated approaches analyze the network of relationships into which a framework is embedded to derive indicators for originality, completeness and significance.

3.4.5.2.3. Integration

Integration functionality needs to establish the links between a framework and existing concepts, theories and methods. When the framework includes novel concepts, it must integrate these concepts into the network of concepts. Integration also requires an analysis of completeness. The evolution of each framework needs to be monitored. Frameworks which are sufficiently matured and which have become the basis for theories may need to be fully converted into a theory.

3.4.6. Data

Data is one of the central elements of research because it provides the link between theoretical concepts and observable phenomena. Because of space constraints, raw data currently cannot be published together with scientific articles.

An alternative are data depositories for unpublished data and data journals which exist, for example, in the chemical field. See Rossmassler (1980, p 68). Information technology does not impose these restrictions and allows publishing the original data together with the findings. This has significant advantages: access to original data may be necessary for judging quality and significance of findings. Moreover, raw data can often be used to extract additional knowledge: published data can be used for further analysis and secondary study by other researchers. More widespread use of secondary data is desirable not only because of the reduction in time and cost but also because reuse of the same data increases scrutiny, thus leading to higher data quality and standardized measures (Jarvenpaa 1990, p 19).

3.4.6.1. Attributes and relationships

Table 29. Attributes of 'data'

Tag	Description
Title	Title (optional)
Description	Short description of the data.
Content	
↳Data files	Actual data values in the form of lists, structured records, tables
↳Meta data	Description of the relationships between elements of the data files
↳Data collection process	Brief description of specific steps taken for collecting and refining the data
↳/Timeline	Lists the main data collection events along a timeline; derived attribute
↳Observers	Authors, experimenters, data collectors
↳Limitations	Description of limitations which other researchers need to be aware of when reusing this data and of any irregularities that occurred.
Type	
↳Quantitative/qualitative	Quantitative / qualitative
↳Objective/subjective	Objective / subjective
Quality	
↳Consistency	Consistency of the data.
↳Completeness	Completeness of data
↳Measurement error	Accuracy of measurement

3.4.6.1.1. Content

Data accompanying a scientific study can be large and complex. This makes it difficult to determine a fixed structure. Nevertheless attributes of data need to fulfill two main needs: they need to describe the context in which data has been created and they need to list the individual pieces including their meaning. Attributes which establish the context are title and description. The former contains a short description of the data and can also be used to uniquely reference the data set. A description attribute should describe the data and the way in which it was collected in more detail. The description needs to include a timeline which lists the main events. The timeline can be maintained as a separate – derived – attribute. In addition, the known limitations of the data need to be listed and the subjects who were questioned or who were involved in data collection need to be named. Any limitations regarding the interpretation of the data or its use by third parties need to be noted. If pseudonyms have been used for the protection of privacy this needs to be noted as well as if additional data has been omitted for the protection of privacy. The limitations attribute is also needed to describe any irregularities which occurred during data collection. Data needs to be related to methods and instruments used for data collection. To characterize the approach for data collection, it is not sufficient to list the general method used (which can be expressed as a relationship between the study from which the data originated and the set of methodologies supported by the infrastructure). Rather, researchers need to describe how the general method has been applied for the specific case (attribute: data collection process). Insofar as processed data is presented, the operations of cleaning and aggregation which have been performed need to be explained.

A second group of attributes is needed to capture the individual data items. Researchers rarely read data directly but prefer to import them into mathematical and statistical software packages which allow analysis and visualization. Standard file formats including tab-delimited files, internet MIME (Multipurpose Internet Mail Extensions, Freed & Moore 1997) encodings, and SGML (Standard Generalized Markup Language (ISO 8879), see also Spring 1991, pp 165f.) have matured sufficiently to reliably extract and move numeric or textual data between many applications on different software platforms. A 'data files' attribute can be used to store large sets of data items. The interpretation of data items and the documentation of relationships between data items is more problematic. Thus another attribute is needed which describes the structure of the data. This includes relationships between tables and lists, possible operations on the fields, as well as number types. Ranges of variables, upper and lower bounds of permissible values, error terms etc. need to be included in the structure attribute. Ideally, data types should be taken from a predefined list of types frequently used in scientific research. Efforts at developing such standards are described by Lerner (1980). Established conventions for describing table specifications in relational or object-oriented databases can be used for the meta structure attribute. When a markup language such as SGML (Standard General Markup Language) or XML (Extended Markup Language) is used, then the raw data and meta structure attributes may be combined.

3.4.6.1.2. Quality

3.4.6.1.2.1. Consistency

Where data can be derived from other data (e.g. as a sum of a column), the calculations must be correct. In addition, data should not contradict established knowledge about values and relationships between variables outside of the phenomena being studied. This second criterion has exceptions. Sometimes measurements which are outside of the expected range indicate a deficiency in the underlying theory.

3.4.6.1.2.2. Completeness

Data needs to be checked for completeness. All elements described in the meta structure need to be present. The range of data needs to be adequate for the data type. In the case of statistical data significant deviations from known distributions need to be addressed.

3.4.6.1.2.3. Measurement error

This attribute records the accuracy of measurement. It is a key criterion for evaluating data.

3.4.6.2. Activities

3.4.6.2.1. Editing and presentation

At present, a variety of tools exist for editing and presenting data. The challenge for a knowledge infrastructure is to incorporate existing tools and to allow for the transformation of data between different tools. This requires standardization of formats and meta structures. The large number of formats and the complexity of data structures present significant hurdles for a practical solution.

3.4.6.2.2. Integration

Data integration is a major problem when the format and the circumstances under which data have been captured differ. A variety of methods for the standardization and extrapolation of data exist which allow to combine data from different sources.

3.4.7. Method

Methods describe how a goal can be attained or a problem can be solved. This includes the procedure or steps to be taken as well as the means and principles to be employed (Chmielewicz 1994, p 36; Miller 1991, p 743; Bunge 1998a, p 8). Descartes characterized this as follows:

> "...by a method I mean certain and simple rules, such that, if a man observes them accurately, he shall never assume what is false as true, and will never spend his mental efforts to no purpose, but will always gradually increase his knowledge and so arrive at a true understanding of all that does not surpass his powers." (Descartes 1927, p 49)

While concepts and theories are passive, methods imply action. They consist of sequences of steps which - if correctly followed - lead to the intended result. Methods need to be testable (Schauenberg 1998, p 47). In many cases, methods are translations of theory into a program of action. Their advantage is to reduce the cognitive load on the user; their disadvantage may be that methods often assume background knowledge on the part of the user that is not explicitly described in the method. In other cases, methods do not follow from theory but have arisen from practice. A goal of research, then, is to substantiate such methods by linking them to theory and by understanding their advantages and limitations. Heuristics as well as current software development methodologies are examples for the latter case. Terms such as 'technique', 'procedure', and 'methodology' are often used as synonyms to method, although the term 'methodology' would more appropriately be interpreted as a theory of methods. Methods are closely related to theories: every method can be reformulated in a theory of action (Laudan 1996, p 134-137). The reverse is not true. Therefore the epistemological issues raised for theories also apply to methods.

While theories may have received more attention than methods in the past, the ability to execute operations automatically on a computer has greatly increased the interest in methods. The often quoted gap between theory and practice is in part the result of an absence of methods with which to apply theories. In a paper-centric world, this gap was partially the result of the scarcity of people who were able to interpret and apply theories methodically. With information technology, it is no longer true that the application of a theory requires reading and assimilating a significant part of the theory. When theories are formulated electronically, they may be executed rather than applied. Thus the gap between theory, methods and practice may shrink significantly in online environments.

3.4.7.1. Attributes and relationships

Table 30. Attributes of 'method'

Tag	Description
Title	Title
Content	
↳ Detail	Detailed description of the approach.
↳ Background	Description of the context of the method
↳ Initial state	Context in which the method can be applied
↳ Resulting state	Result and objective of the method
↳ Limitations	Description of limitations which other researchers need to be aware of when using this method and of known irregularities that may occur.
Type	
↳ Degree of formalization	Distinguishes automated methods from methods requiring considerable adaptation and interpretation
Quality	
↳Accuracy	Measure of exactness
↳Precision	Degree of repeatability
↳Robustness	Capacity to main unaffected by small variations in parameters

3.4.7.1.1. Content

Methods can be reformulated as theories. Therefore attributes and activities which are relevant for theories should also be relevant for methods. Methods have all content attributes of theories. The content attribute of a method needs to consist of at least four parts:

3.4.7.1.1.1. Detail

Similarly to the detail attribute of theory and concept, this attribute allows the verbal description of the network of relationships in which a method is embedded.

3.4.7.1.1.2. Background

This attribute describes the context in which a method has been developed.

3.4.7.1.1.3. Initial state

This attribute describes the situation(s) in which a method can be applied. It includes the context in which a method can be applied and mechanisms for distinguishing between suitable and unsuitable contexts, if available. The initial state may be described via statements and thus can be a derived attribute.

3.4.7.1.1.4. Resulting state

This attribute describes the goal of the methods. The resulting state may consist of several alternative states and may include states which are caused by errors during the execution of a method. The resulting state may be described via statements and thus can be a derived attribute.

3.4.7.1.1.5. Limitations

This attributes describes any limitations of the method. Methods may occasionally lead to incorrect or spurious results (irregularities). The purpose of this attribute is to capture all problems which may occur in the application of a method.

The boundary between methods and theories is not sharp. The more unambiguously a method is defined, the closer it may resemble a causal theory. However, it is not useful to merge both concepts into one. Methods often describe efficient operationalizations of theories. They are thus subordinate to theories. In addition, methods often exist which – although being practically useful – are neither unambiguously described nor firmly established from a theoretical perspective. Information systems development methodologies provide are a rich source of examples for the latter (for an overview of attempts to increase the formalization of such methods see Brinkkemper, Lyytinen & Welke 1996). Similarly to most other elements of the model of scientific knowledge, methods need to be related to concepts on which they operate. They are frequently the source or solution of problems.

3.4.7.1.2. Type

Methods differ from theories in the type attribute. Theories can be distinguished according to their analytical focus and their scientific goal. By their very nature, however, methods always have an analytical focus on outcome. Similarly, the scientific goal of methods favors prediction over understanding. A method needs to reliably lead to the stated objective. Therefore analytical focus and scientific goal are not suitable differentiators for methods.

3.4.7.1.2.1. Degree of formalization

Methods vary in their degree of formalization. The most highly formalized methods are implemented in software and require no human intervention for guidance or decision making during their execution. Algorithms which are unambiguously specified and require no additional decisions for operationalization are high in formalization. In contrast procedures and guidelines which need considerable interpretation and customization by the user are low in formalization.

3.4.7.1.3. Quality

In the physical and biological sciences, standards have been developed for analyzing the performance of methods. In pharmaceutical research, several methods are used for evaluating the performance of analytical methods. At least three of these measures can be generalized beyond analytical methods:

3.4.7.1.3.1. Accuracy

"Accuracy is the measure of exactness or the closeness of agreement" (Swartz & Krull 1997, pp 56-67) between the achieved results of the method and the values that are accepted either as conventional, correct or reference outcomes.

3.4.7.1.3.2. Precision

"Precision is the measure of the degree of repeatability of a" ... "method under normal operation and is normally expressed as the percent relative standard deviation for a statistically significant number of samples." (Swartz & Krull 1997, p 57)

3.4.7.1.3.3. Robustness

"Robustness is the capacity of a method to remain unaffected by small deliberate variations in method parameters" (Swartz & Krull 1997, p 67).

3.4.7.2. Activities

3.4.7.2.1. Editing and presentation

Similarly to theories, tools are needed for creating and visualizing methods. Process models as defined in information systems development methodologies such as ARIS (Scheer 1998a) can be used to graphically depict the sequence of steps. Textual descriptions in structured or unstructured formats are another alternative.

3.4.7.2.2. Integration

Like theories, methods need to be linked into the existing body of scientific knowledge. The problems which arise and the approaches which can be used are similar to those described for theories in section 3.4.3.2.2. on page 165 above.

3.4.7.2.3. Application

An important aspect of methods is their application in the course of research. Application services need to support the adaptation of a method to a specific research task, the identification of methodological choices and problems, and tracking the progress of the method. Ideally, methods could be instantiated in the form of software which would guide a researcher. Such software would also create a research protocol and could store and organize data which is being collected.

3.4.8. Project

Scientific knowledge arises from research projects and research studies. Information about these projects is often needed for interpreting theories, methods and concepts. Therefore online infrastructures need to capture information about individual projects in a separate class. Much of the information about projects is reflected in relationships: a study may be based on existing problems and theories; it may adopt a specific method and generate data and facts. This in turn may lead to novel scientific ideas which are based on the project.

3.4.8.1. Attributes and relationships

Table 31 lists the attributes of projects, most of which speak for themselves.

Table 31. Attributes of 'project'

Tag	Description
Title	Title of the project
Description	Short description; summary of the project
Content	
↳Objectives	Objectives of the project
↳Background	Reason for and justification of the project
↳Research design	Approach chosen for the project and rationale
↳Detail	Detailed description of the content and the results of the project
↳Duration	Time frame in which project was conducted
↳Limitations	Description of limitations of any irregularities that occurred.
Type	
↳Type of objective	Exploratory, descriptive or explanatory
↳Degree of control	Whether researcher has control over variables in an empirical study (experiment)
↳Research environment	Whether research is conducted in field or laboratory
↳Time dimension	Longitudinal or cross-sectional
Quality	
↳/Replicated by	List of studies which have replicated this study

3.4.8.1.1. Content

3.4.8.1.1.1. Objectives

This property describes the scientific objectives of a study. They may range from the confirmation or rejection of a hypothesis to the development of a theory or the collection of data. Objectives belong to one of three types of objectives, which are described further below in the section on 'types': descriptive, predictive or exploratory. The description of objectives should include an indication of the expected significance and utility of the expected findings. Note that not all scientists agree on the need to select valuable research problems and therefore might object to the latter objective.

3.4.8.1.1.2. Background

This attribute describes the context in which the study originated. It describes the participants involved in the project and additional information which is necessary for understanding the context of the study that is not contained in one of the other attributes.

3.4.8.1.1.3. Research design

The research design describes the approach selected for solving the research problem and the rationale for selecting this approach. The approach selected will often be an instance of one or several methods which are already described in the methods section of the knowledge infrastructure. In addition, the research design should identify alternatives to the selected approach and outline their disadvantages.

3.4.8.1.1.4. Results

This attribute captures a verbal summary of the project's results. The detailed results are reflected in the hypotheses which have been confirmed or rejected by the data generated as well as the theories, methods, problems etc. which are formulated on the basis of the project. The results should be ordered by their significance. They should not be limited to positive results but should also include the failure to confirm important hypotheses. In addition, when misinterpretations of results are likely, they should be explicitly addressed in this attribute.

3.4.8.1.1.5. Duration

This attribute describes the time frame during which the study was performed and its detailed temporal characteristics. The distinction between longitudinal and cross-sectional analysis is also reflected in summary form in the 'time dimension' attribute of the 'type' section.

3.4.8.1.1.6. Limitations

This attribute captures limitations of the project which the reader needs to be aware of. It needs to mention constraints and irregularities which affected the execution of the project.

3.4.8.1.2. Type

The detailed information contained in the attributes and relationships of projects allows to distinguish different types of projects. The categories are useful for searching the knowledge infrastructure, for determining similarities between different projects and for selecting and evaluating the methods used in a project. The following categories need to be distinguished:

3.4.8.1.2.1. Type of objective

Three main types of research objectives need to be distinguished: exploratory, descriptive and explanatory. Exploratory projects are conducted in the initial phases of research to become familiar with a phenomenon. They provide a starting point for theory development and for formal empirical studies. An example for approaches used for this type of projects are focus groups (Babbie 1995, p 84). Descriptive research projects describe situations and events concerning the phenomena of interest. Their emphasis is on observation and the accurate collection of data. Explanatory research projects construct or test theories. Research projects may combine several of these objectives.

3.4.8.1.2.2. Degree of control

This attribute refers to the ability of the researcher to manipulate variables in an empirical study. In an experiment – whether in the laboratory or in the field – the researcher exerts active control over key variables. This allows meticulous testing of the relationships between different variables. In other modes of research such as survey and field research, scientist avoid exerting influence.

3.4.8.1.2.3. Research environment

"Research designs also differ on whether they take place under actual environmental conditions or under artificial or simulated conditions. These are called field and laboratory studies, respectively. The main distinction hinges on whether the study is of subjects under normal conditions for the problem studied." (Emory 1985, p 61)

3.4.8.1.2.4. Time dimension

This attributes is used to distinguish between longitudinal studies and studies that focus on a single point in time.

3.4.8.1.3. Quality

Three quality criteria need to be distinguished for empirical research projects. Internal validity is violated if the conclusions drawn from the research do not accurately reflect the phenomenon under study. External validity concerns problems of generalizing the findings across persons, settings and times (Campbell & Stanley 1963, p.5; Cook & Campbell 1979, pp. 51-55). Studies increase in importance when their results can be replicated by other researchers. This can be captured by the 'replicated by' attribute. This is an attribute which can be derived from an analysis of the network of knowledge elements. A replicating study will be linked to the original study via a based-on relationship. This allows automated algorithms to compare the findings in both studies.

3.4.8.2. Activities

3.4.8.2.1. Editing and presentation

Research projects tie together many different elements of knowledge. Therefore a key challenge for specialized editors is to simplify the creation of links between the theories, problems, methods, hypotheses, data etc. which is at the core of a project. A major challenge is the development of an intuitive diagrammatic representation. The verbal description of research projects may be simplified by providing type-specific templates.

3.4.8.2.2. Integration

Projects need to be linked to many elements of knowledge. The approach taken in a project can be recorded as part of the methods class. When a corresponding method does not yet exist, a new method entry needs to be added to the knowledge infrastructure. Empirical research projects also need to be linked to the data which has been collected. Via statements and methods, projects can be linked to theories and concepts. A direct link may exist between a project and the problem which it tries to solve or which arises from it.

4. Implications

4.1. Feasibility: IS Cybrarium

The previous chapters have shown that significant change is possible for scientific knowledge infrastructures. Nevertheless, many practical obstacles may remain for the evolution of scientific knowledge infrastructures. Current technology lacks maturity; the scientific community may be unwilling to adapt. Moreover, some opportunities for leveraging information technology and some of its impacts may have been overlooked.

Additional insights about the potential evolution of scientific knowledge infrastructures can be gained by experimenting with an actual online knowledge infrastructure. For this purpose, the author has developed an internet-based infrastructure which is targeted at scientific knowledge in the field of information systems. The system is called 'Information Systems Cybrarium.' It is accessible at http://cybrarium.usc.edu. The system has also been developed to demonstrate the technological viability of the concepts described in the previous sections, to experiment with alternative implementation paths, to identify additional opportunities for leveraging information technology and as an experimentation platform to gain an understanding of user acceptance and preferences.

The Cybrarium is a collaboratory. It provides extensive support for the creation and synthesis of knowledge as well as for its dissemination. Participants can create contributions, annotate existing contributions and publish either instantly. Figure 26 shows how the Cybrarium is integrated into the research environment. Besides scientists, its main stakeholders are reviewers, students and practitioners. Each of these groups has different goals in using the Cybrarium. Practitioners, for example, are more concerned with applying scientific findings and identifying and prioritizing problems. Students need to explore the knowledge base and systematically acquire knowledge while scientists strive to extend the current body of knowledge, evaluate new contributions and collaboratively develop new solutions. As the figure shows, the Cybrarium is linked to core scientific journals in the field of information systems and incorporates their abstracts and tables of contents. The Cybrarium is an open infrastructure; it can incorporate articles which are stored on any external site. Its homepage is shown in Figure 27.

Technically, the Cybrarium is powered by a relational database and a custom-developed application which interfaces with a standard web-server. The architectures will be described in more detail further below. The database implements part of the model of scientific knowledge developed above. It distinguishes different categories of knowledge units. Examples are a 'concept' category which includes concepts and theories. A 'text' category represents the most generic unit of knowledge. The meta structure also contains a 'problem' category which is needed for the identification and prioritization of research problems. The Cybrarium provides additional functionality for collaborative filtering which enables readers to retrieve knowledge according to their personal review function, annotation and multi-lingual user interfaces which will be further discussed below.

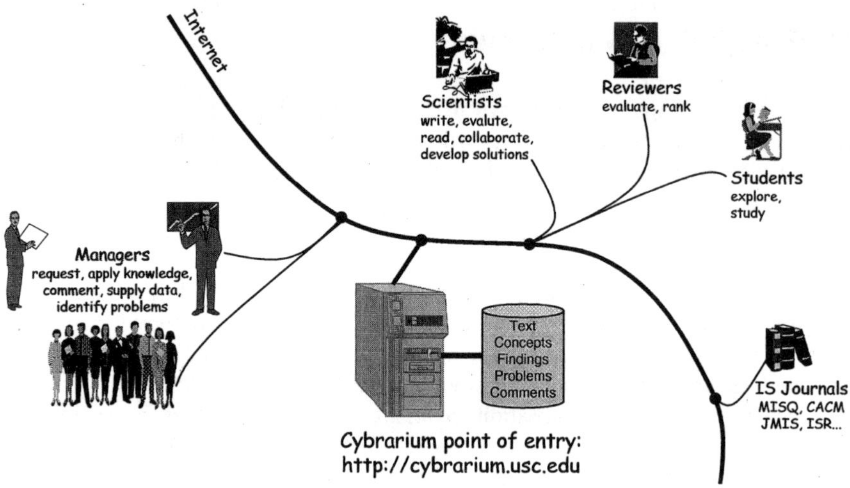

Figure 26. Information Systems Cybrarium

The Cybrarium has been under development since 1996. The current web-based system is the fourth prototype developed at the Marshall School of Business, University of Southern California. It originated in a Ph.D. seminar on "Information Systems Research in the 21st Century" (Hars et al. 1999). The first three prototypes were standalone systems which were not accessible from the internet.

In the following the main features of the Cybrarium that have an impact on the way that scientific knowledge is managed or on the research process will be described.

4.1.1. Cybrarium architecture

The current architecture of the Cybrarium infrastructure is shown in Figure 28. It uses a thin client-server architecture. The clients require only a web browser; the server houses a relational database and the application logic. A second server is used for monitoring the Cybrarium server and for generating email messages.

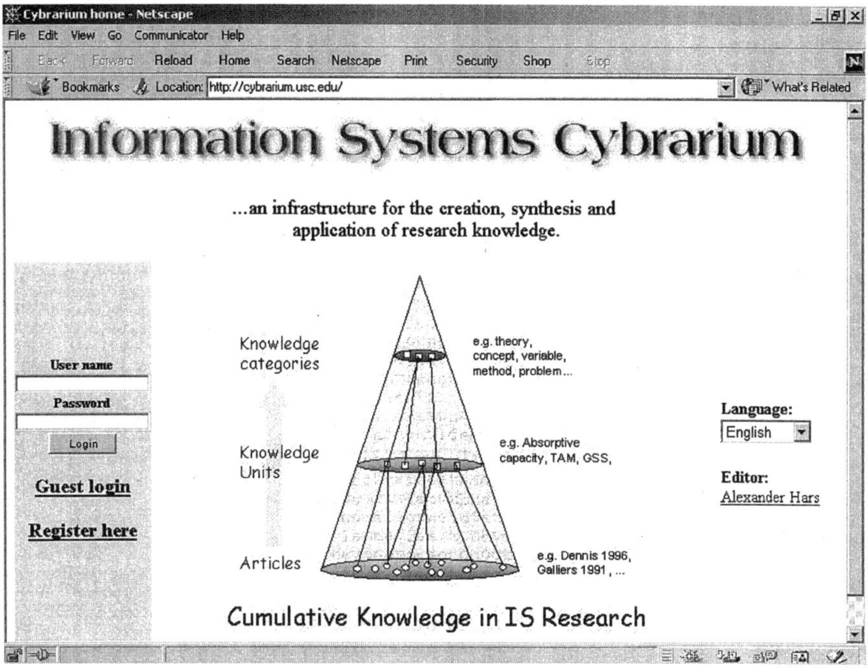

Figure 27. Cybrarium screen shot

In the following the operation of the Cybrarium will be described using a simple session as an example. A user initiates a Cybrarium session by accessing the Cybrarium home page at http://cybrarium.usc.edu from his browser (1). The user then chooses between different logon options which range from anonymous access to new user registration. The logon request is received by the Internet server (2) running on the Cybrarium server. It is then passed on to an off-the shelf interface (3) which connects the internet server with the main Cybrarium application (4). The interface is necessary because the main application is written in Object Pascal using the Delphi environment for rapid application development. The main application analyzes the requests, accesses the relational database (5) which stores data, user profiles etc. If possible, it determines the user. If the user is new and the request contains his registration data, then the new user's data is written to the database. Next, the registration manager (6) generates an email message which confirms the username and password. The user receives this message on his email client (7) and must respond to the message for authentication. If no reply is received within a period of 48 hours, the account is deactivated. Readers are not required to enter username and password every time. These parameters can be stored in an HTTP cookie (8). After logon, the user may retrieve content stored in the Cybrarium, execute searches, add new knowledge, create comments etc. These activities follow the same basic sequence where the request is passed via the interface to the main application which reads data from the database or writes new content to the database and then generates a response which is transferred to the client's browser.

Two Java applets provide additional functionality on the client side. One applet can be used as context-sensitive menu (9). Another Java applet functions as diagram editor (10). It can be used to graphically represent knowledge structures and to navigate through these structures. The diagrams are stored in the Cybrarium database. Access is provided via Java's Remote Method Invocation interface which allows the communication between the applet as client and a Java servlet on the Cybrarium server (11).

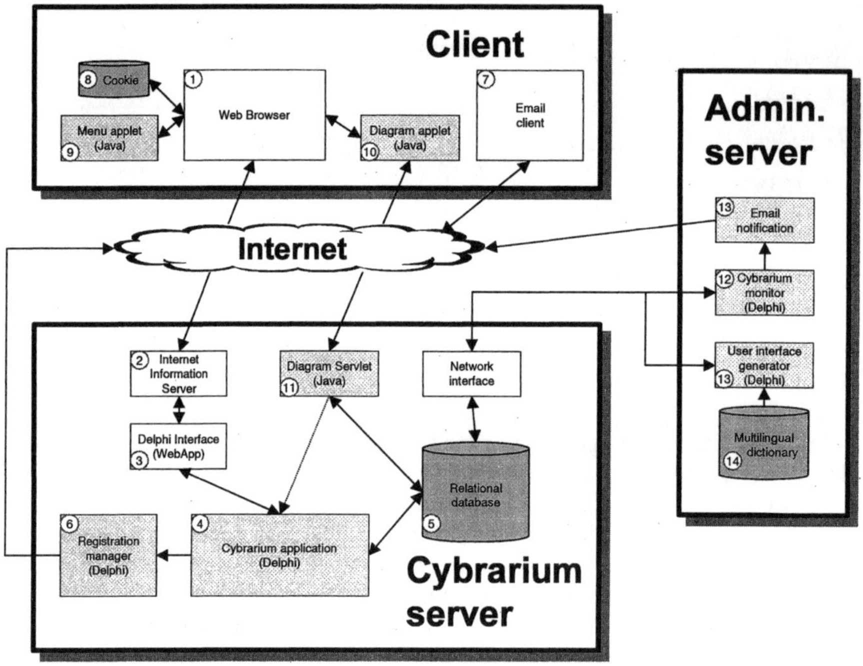

Figure 28. Cybrarium architecture

Additional functionality is provided by a separate administration server. In regular intervals, a monitoring program (12) verifies that the Cybrarium server remains active. If the server fails to respond the Cybrarium administrator is alerted via email. Every night, the monitoring program prepares a summary of the previous days' activities. In addition it notifies authors who have received comments to their contributions. The monitoring system also alerts readers about new articles which they may be interested in. Readers can request notification by topic or author. Another Cybrarium module (13) generates the multi-lingual user interface. It uses a custom dictionary to add language specific text into html pages and templates used in the Cybrarium.

4.1.2. Meta structure

A core difference between the Cybrarium and other scientific knowledge infrastructures is its meta structure. While most electronic journals adopt the traditional structure of volumes and articles, the Cybrarium is based on part of the model of scientific knowledge described above (see Figure 18 on page 121). Thus when an author adds an article to the Cybrarium, he needs to identify the content type. Traditional articles, which are collections of many thoughts and insights, are classified as general texts. Knowledge, however, can also be contributed in the form of theories, definitions of concepts, variables, problems, data etc. The Cybrarium, therefore, encourages a more precise specification of knowledge. While a single paper may contain many different insights, the author is encouraged to separate his contributions into knowledge about specific theories, variables or problems etc. This enables a more precise - even though not necessarily mathematical - formulation of knowledge and a higher degree of integration between contributions of different authors.

4.1.3. Open architecture

Contributions to the Cybrarium can be made either by adding the full text to the Cybrarium database or by submitting a link to the publicly accessible URL where the text is permanently available. The Cybrarium, therefore, can incorporate any html-page available on the internet. This has the advantage of reducing resource requirements for the Cybrarium; more importantly, it simplifies extending the knowledge available in the Cybrarium. In addition it reduces the effort for an author to contribute his findings to the Cybrarium. Many of his texts may be incorporated by default. Thus the Cybrarium fulfils the function of an index server, which guides readers to his contribution. Copyright issues or specific publishing formats do not stand in the way, as the articles are incorporated into the Cybrarium by reference only.

Technically, the open structure of the Cybrarium is implemented as follows: Similar to every document in the Cybrarium, each external html page receives a unique Cybrarium identifier. When this page is requested from a reader, the external html page is displayed in one frame (see Figure 29). The second frame provides access to Cybrarium services which allow to view and create comments, evaluate or review the article, identify links and relationships etc.

Information technology enables knowledge infrastructures to focus on the integration and accumulation of knowledge for a domain. When several competing sites and online journals exist which publish such knowledge, this goal can only be achieved if the infrastructure is open. It must be able to incorporate materials from other sites.

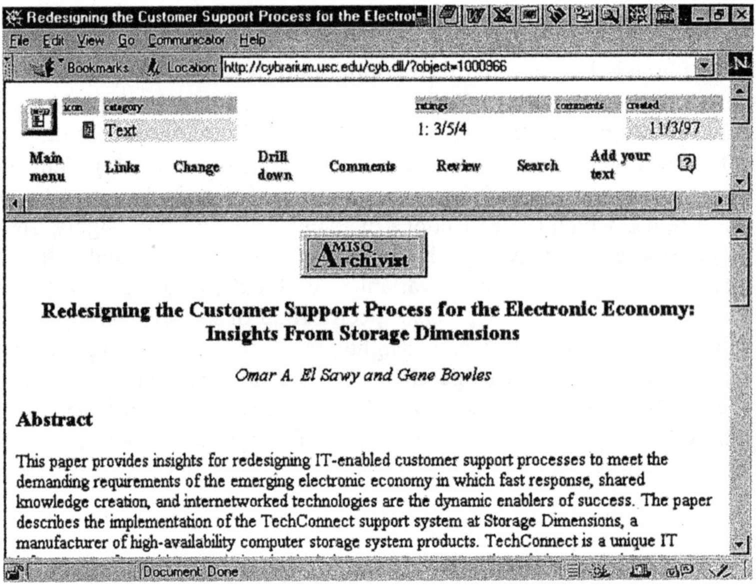

Figure 29. An article from MISQ viewed from within the Cybrarium

4.1.4. Meta journal

The open nature of the Cybrarium described in the previous section enables the incorporation of articles which are published in other journals on the internet. The publishers of paper and electronic journals increasingly provide access to abstracts free of charge. In some cases, this applies to the whole article. In either case, the source abstract or article can be imported via an html reference into the Cybrarium. Since January 1998, a team of 6 faculty and PhD students has indexed the major four top journals in Information Systems in this way. While a reader can browse these journals within the Cybrarium by issue and volume, he has the added benefit of being able to access articles from different journals for the topics which he is interested in.

Thus the Cybrarium is becoming a journal of journals – a meta journal. A consequence of this approach will be that in the future the same article may appear in several virtual or meta journals. This is desirable as it indicates that the article appeals to the combined audiences of several journals.

4.1.5. Collaborative filtering

Information technology eliminates many of the barriers for the dissemination of knowledge. The costs of publishing are nearly zero, the time lag for publishing has nearly been eliminated and knowledge is accessible from anywhere in the world with internet access. With the wealth of information thus made available, researchers face the problem of information overload. It becomes almost impossible to keep up to date with the advances in the general field. Top-tier journals have been one answer to this problem: they promise to identify contributions with high significance to the overall field. Their disadvantage is a lengthy review process and the inability to focus on advances in highly specialized sub-fields.

In an electronic environment additional approaches for the identification of promising contributions are possible. In contrast to paper technology, immediate feedback from the reader is possible. In the Cybrarium every node is displayed together with a streamlined feedback form which requires a maximum of three mouse-clicks. The reader is encouraged to rate a paper according to three dimensions: level of detail, rigor and relevance. On each dimension 7 point Likert scales are used. 'Level of detail' is used to assess the amount and depth of knowledge conveyed in the contribution. 'Relevance' helps to identify the value and usefulness of the contribution. It is especially important for practitioners. 'Rigor' refers to the "academic" qualities of the paper. Its foundation in existing theories, the soundness of its measures and the validity of its arguments. The three variables have been chosen to reflect concerns about the quantity of contribution and of its value from a practical and scientific perspective. However, it would be possible to argue for additional or other measures. For example, the originality of a contribution might provide additional insights. The larger the number of variables, however, the higher is the effort of the reader for evaluation and the lower is the number of responses.

In the Cybrarium, average ratings for all three dimensions are displayed for each contribution together with the total number of ratings. Every reader can create filters which eliminate contributions that are below reader-defined thresholds. A student who is interested in a general overview about a domain, may elect to view only those contributions which are rated exceptionally high in relevance and rigor. A researcher who is interested in the application of specific methodologies may target only papers which are high in rigor. A researcher who wants to keep abreast of avenues pursued by his colleagues, however, may need to view all papers in his research area.

The rating mechanism depends upon active participation by all readers. Therefore the system keeps track of a reader's rating behavior and informs readers when they rate too few papers. The system is also able to identify readers who consistently rank contributions too low or too high. Further experience will show whether the analysis of rating behavior is sufficient to counteract attempts for manipulating the filters.

4.1.6. Annotation

An important element of research is discourse among scientists. Information technology allows to add annotation features to a knowledge base. In the Cybrarium, any registered user can add comments to any contribution. As every comment is treated as a standard contribution, it is possible to comment on comments and to rate the quality of comments. In addition, the Cybrarium can automatically notify an author that one of her contributions has received a comment. This notification feature can be disabled by the author.

Currently annotation targets a contribution as a whole. It would be desirable to add more fine-grained annotations which enables a reader to comment on a specific sentence, paragraph or figure.

4.1.7. Notification

Knowledge infrastructures such as the Cybrarium operate in an asynchronous mode. Collaborators submit, read and evaluate contributions at different times. While such a setup is flexible, cases arise when a tighter coupling between the partners is needed. The Cybrarium therefore contains an email notification mechanism. A participant is automatically notified when one of her contributions has been commented upon. This potentially opens a channel for dialogue with the reviewer. Should the participant reply to the comment a similar notification is sent etc. Currently the mechanism is being extended to alert a participant about contributions in a topic of interest.

4.1.8. Instant publishing

In most paper-based and electronic knowledge bases, contributions are reviewed before they are incorporated. This has a significant disadvantage: There usually is a significant time-lag between submission and publication. While information technology can reduce some of the delays (e.g. transmission times) other significant sources of delay remain. Examples are coordination problems and the need for slack to accommodate reviewers' other commitments.

In the Cybrarium, the lead time from submission to publication has been reduced to zero. Contributions are published instantly. As we have argued above this is not as radical a change as it might seem: While until the Eighties, researchers needed paper journals to make their insights accessible to a large number of their colleagues, the internet now enables them to publish their results instantly by themselves. This does not guarantee readership. Similarly, if a contribution is added to the Cybrarium, this does not automatically lead to a large number of readers. Readers may have set their filters to exclude contributions which have not yet been evaluated. Others may not be interested in the topic. However, readers with a special interest in the topic will immediately have access to the contribution. Some of them will rank the paper. The higher a paper is ranked, the more likely it is that the paper will be read by a wider audience. In addition, lists of highly recommended contributions can be maintained by individuals or groups within the Cybrarium. These lists may take the place of journals in directing the attention of their audience to papers of high quality for a specific domain. In the Cybrarium, therefore, submission and review (or recommendation) have been separated. Review can be performed after publication.

4.1.9. Competitive contributions

One of the fundamental principles in science is the competition of ideas. As knowledge in a domain matures, contributions which have previously been held in high esteem may be disputed and finally discarded. In a paper environment, an explicit record of accepted and disputed or discarded knowledge is not kept. In an electronic environment which aims to provide access to the latest knowledge in a domain the trail of contributions can be kept. In the Cybrarium, contributions are never discarded. Any text remains accessible to the system unless it is voluntarily removed by its author. However, the combination of evaluation and filtering mechanisms described above allows to limit access only to those contributions that are up-to date.

A different problem, however, arises for those contributions which reflect summaries and syntheses of other contributions. 'Concept' nodes, for example, are intended to summarize the knowledge reflecting a specific concept. Such nodes are much less stable than individual texts, because they need to reflect many small advances in a topic. In the Cybrarium, therefore, a mechanism is being implemented which allows to update descriptions of general scientific concepts without prior consent of the previous author.

When a researcher wants to extend or modify the description of a concept such as 'systems development life-cycle' or 'electronic meeting system', he may instantly replace the current version. The previous version is not discarded; a link is established from the new version to indicate that it has become its successor. The new author provides a description of the changes, which is used for a history of the concept. When a new reader accesses the concept, the new version will be displayed automatically. To ensure quality, a mechanism needs to be embedded which prevents researchers from lowering the quality of the concept description. Therefore, the author of the previous version is automatically alerted via email and asked to review the new version. If he accepts the changes, the new version permanently replaces the previous version. If he does not agree with the new version, then the conflict is escalated. A group of reviewers is determined via an algorithm which is partly based on random selection and partly based on an analysis of links as an indicator of interest in the subject. The reviewers are asked to compare both contributions and then to vote in favor of one or the other. After a period of one week for extensions and four weeks for significant modifications, the version with a majority of votes is selected. A tie is resolved in favor of the previous version.

Different types of competitive modifications need to be distinguished. An extension of the previous version which leaves previous statements unchanged but adds new knowledge is generally less problematic than a substantial revision which changes key elements of the previous texts. In addition, authorship becomes a complicated issue. Who is to be regarded as the author of knowledge which is derived from a large body of contributions and which is continuously updated? Several alternatives exist: As it is difficult to compare the relative contribution of each author, one might argue that no author should be shown. The disincentive, however, would be too large. When all authors are listed, the sequence of authors becomes important. Should the author who first wrote about the concept always be listed first? Or should the reverse order be used - the author of the most recent version would always be displayed first. For incentive reasons the latter is preferable. Otherwise, an author who would contribute a new version might find herself in the same situation as when no author would be listed. In addition, among a community of researchers who are collaborating in the field, it is especially important to be aware who is currently working on the concept.

4.1.10. Identification of promising research problems

The role of traditional journals has been to disseminate research results. They are targeted at the final stages of the research process. They certainly have an impact on the early phases by setting expectations about what constitutes acceptable results, what kinds of approaches need to be used and by proposing areas for future research. Because of the inherent limitations of paper to support fruitful discussions, however, they have not focused on the identification of research problems and the selection of priorities.

This is a significant shortcoming. It is widely recognized that the identification of a worthy research problem may be a major hurdle in the research process. In some cases, the clear identification of a problem is the most crucial step in a research project. In addition, criticism from industry that academia focuses too much on "toy problems" and is not sufficiently aligned with the realities, reflects a fundamental disagreement about what kinds of problems need to be studied. An improvement of research productivity, therefore, may be reached by ensuring that scientists pursue promising and valuable research problems. This does not imply that science should be driven by potentially short-term interests from industry. In contrary, it implies that scientists need to increase their attention about the relevance of their research to the body of scientific knowledge

The Cybrarium provides the ability to define problems and to relate these problems to subject areas, variables and theories. Senior researchers and representatives from industry can comment on the relevance of problems and the quality of problem descriptions. The Cybrarium provides a list of most pressing problems identified in the IS area which can be consulted by funding agencies, by PhD students and by any researcher looking for a way to maximize their impact.

4.2. Conclusion

The previous chapters have shown that information technology may fundamentally change the intellectual infrastructure for the sciences. Information technology changes fundamental characteristics of knowledge, such as the ability to update published knowledge, to execute knowledge in the form of software and to adapt knowledge dynamically to the needs of the reader.

As a consequence, online knowledge infrastructures are not likely to replicate traditional paper-based journals. They will be much more interactive and improve discourse among researchers. They will not just focus on articles but break knowledge down into relevant components such as theories, methods, data, problems etc. At the same time, they will include functionality for assembling these components dynamically into custom articles which are targeted for the needs of the individual reader. Online knowledge infrastructure thus will combine characteristics of many current infrastructures, including collaboratories, logic servers, encyclopedias, digital libraries and electronic journals.

The most important change, however, may be the change in focus. While current journals aim at the dissemination of scientific knowledge, future online infrastructures may turn the focus more towards the creation and integration of scientific knowledge. This has important implications for scientific processes. The body of knowledge no longer will be a loosely connected sea of individual contributions but scientists will have the opportunity of adding their knowledge directly into the contributions of other scientists. While such integration poses few technical problems, many procedural and practical issues – particularly the impact on the scientific reward system and on the ownership of ideas – still need to be worked out.

Like many socio-technical systems, scientific knowledge infrastructures are not natural phenomena. Their evolution depends on the choices made by the community of scientists and by the other stakeholders. One of the critical issues for the development of these infrastructures is the current fragmentation of scientific knowledge across many scientific journals. Although information technology provides the means for establishing an integrated, even if distributed body of scientific knowledge, standards are needed for linking contributions across knowledge infrastructures. The model of scientific knowledge developed in chapter 3 can provide a foundation for such an integrating framework.

The evolution of online scientific infrastructures is only in its early phases. The preceding analysis has shown that the potentials of improvement are large and that the practice of science will be transformed significantly. It is in the hands of the scientific communities to ensure that the evolution proceeds towards an integrated, highly responsive, ubiquitously accessible body of scientific knowledge where scientific processes and information technology are optimally aligned.

References

Agger, B. (1989): Socio(onto)logy – A disciplinary reading. Urbana, IL (Univ. of Illinois Press).

Aiken, M.W.; Sheng, O.R.; Vogel, D.R. (1991): Integrating Expert Systems with Group Decision Support Systems. In: ACM Trans. Inf. Syst. (9:1), 1991, pp 75-95.

Alford, R.R. (1998): The craft of inquiry. New York (Oxford Univ. Press).

Armstrong, T.S. (1997): Peer review for journals: Evidence on quality control, fairness and innovation. In: Science and Engineering Ethics, vol. 3, pp 63-84.

Arnold, K. (1993): The scholarly monograph is dead. Long live the scholarly monograph. In: Okerson (1993), pp 73-79.

Babbie, E.R. (1994): The practice of social research. 7th ed, Belmont (Wadsworth).

Babbitt, T.G. (1997): Operational Definitions in MIS Theory Building. In: J.D. Gupta (ed.): Proceedings of the 3rd AIS Americas Conference on Information Systems 1997, pp 988-990 (http://hsb.baylor .edu/ramsower /ais.ac.97/papers/babbitt.htm).

Bacharach, S.B. (1989): Organizational theories: Some criteria for evaluation. In: Academy of managemement review, vol. 14, no.4, pp 496-515.

Bar-Hillel, Y. (1970): Aspects of language. Amsterdam

Barua, A.; Chellappa, R.; Whinston, A.B. (1995): Creating a collaboratory in cyberspace: Theoretical foundations and implementation. In: Journal of Organizational Computing, vol. 5, no.4, pp 417-442.

Bavink, B. (1932): The anatomy of science. London.

Benson, D. A. et al. (1998): GenBank. In: Nucleic Acids Research, vol. 26, no. 1, pp 1-7.

Berger, J.; Zelditch, M. (1993): Orienting strategies and theory growth. In: Berger, J.; Zelditch, M. (eds.): Theoretical research programs: Studies in the growth of theory. Stanford (Stanford Univ. Press), pp 3-19.

Bergmann, G.(1957): Philosophy of science. Madison (Univ. of Wisconsin Press).

Bernstein, R.J. (1985) (ed.): Habermas and modernity. Cambridge (Polity Press), p 10.

Bernzen, R. (1990): Modell. In: Sandkühler, H.-J. (ed.): Europäische Enzyklopedie zu Philosophie und Wissenschaften. Band 3. Hamburg, pp 425-432.

Beyer, J.M. (1982): Explaining an unsurprising demonstration: High rejection rates and scarcity of space. Commentary. In: Harnad (1982), pp 18-19.

Beyer, J.M. (1978): Editorial policies and practices among leading journals in four scientific fields. In: Sociological Quarterly, vol. 19, pp 68-88.

Bierly, P.; Chakrabarti, A. (1996): Generic knowledge strategies in the US pharmaceutical industry. In: Strategic Management Journal, Vol. 17, Winter 1996, pp 123-135.

Bierman, A.K.; Gould, J.A. (1981): Philosophy for a new generation. 4th ed. New York (Macmillan).

Bizjak, P. (2000): Mankind's Memory Managers: A New Paradigm of Library Science. In: Library Philosophy and Practice, vol. 2, no. 2. http://www.uidaho.edu/~mbolin/bizjak.html (3/4/2001).

Booch, G.; Jacobson, I.; Rumbaugh, J. (1998): The Unified Modeling Language User Guide. Reading (Addison Wesley).

Boland, J. (1994): Designing information technology for distributed cognition. In: Org. Science, Aug. 1994.

Bornstein, R.F. (1990): Publication politics, experimenter bias and the replication process in social science research. In: Journal of social behavior and personality, vol. 5, pp 57-81.

Borst, P.; Akkermans, H.; Top, J. (1997): Engineering Ontologies. In: Int. Journal Of Human-Computer Studies, (46:2-3), Feb-Mar, pp 365-406.

Boulding, K.E. (1956): General systems theory – the skeleton of science. In: Management Science, vol. 2, no. 3.

Braman, S. (1995): Scholarly publishing in the information economy. In: Okerson (1995), pp 39-49.

Bridgeman, P. (1991): The operational character of scientific concepts. In: Boyd, Gasper, Trout (eds.) (1991), pp 57-69. Reprinted from: Bridgeman, P. (1927): The logic of modern physics, chapter 1, pp 1-32. Macmillan.

Brin, S.; Page, L. (1998): The Anatomy of a Large-Scale Hypertextual Web Search Engine. http://google.stanford.edu/~backrub/google.html (8/28/1998)

Brinkkemper, S.; Lyytinen, K.; Welke, R.J. (eds.) (1996): Method engineering – Principles of method construction and tool support. London (Chapman & Hall).

Brittanica Online (1998): Version 98.2. http://www.eb.com (9/14/1998).

Buckingham Shum, S.; Motta, E.; Domingue, J. (2000): ScholOnto: An ontology-based digital library server for research documents and discourse. In: Int. Journal on Digital Libraries, vol. 3, no. 3 (August), pp.237-248. http://kmi.open.ac.uk/projects/scholonto/docs/ScholOnto-IJoDL-2000.pdf (4/9/2001)

Bunge, M. (1998a): Philosophy of science (Revised edition of 'Scientific research' (1967)), vol. 1: From problem to theory. New Brunswick (Transaction Publ.).

Bunge, M. (1998b): Philosophy of science (Revised edition of 'Scientific research' (1967)), vol. 2: From explanation to justification. New Brunswick (Transaction Publ.).

Bunge, M. (1979): Treatise on basic philosophy, vol. 4: Ontology II: A world of systems. Dordrecht (Reidel).

Burks, C. et al. (1990): GenBank: Current status and future directions. In: Methods in enzymology, vol. 183, pp 3-22. SCIENCE QP601.C733

Bush, V.: As we may think. In: Atlantic Monthly, vol. 176, no. 1 (July 1945), pp 101-108. http://www.theatlantic.com/unbound/flashbks/computer/bushf.htm

Butler, H.J. (1995): Research into the reward system of scholarly publishing. Where does scholarly electronic publishing get you? In: Okerson (1995), pp 167-177.

Campanario, J.M. (1998a): Peer review for journals as it stands today – Part 1. In: Science Communication. Vol. 19, no. 3, March 1998, pp 181-211.

Campanario, J.M. (1998b): Peer review for journals as it stands today – Part 2. In: Science Communication. Vol. 19, no. 4, June 1998, pp 277-306.

Campbell, D.T.; Stanley, J.C. (1963): Experimental and quasi-experimental designs for research. Chicago (Rand McNally).

Case, D.O.; Higgins, G.M. (2000): How can we investigate citation behavior? - A study of reasons for citing literature in communication. In: JASIS vol. 51, no. 7, pp 635-645.

Casti, John L. (1995): Complexification. New York (Harper).

Chalmers, A. (1990): Science and its fabrication. Minneapolis (Univ. of Minnesota Press).

Chaitin, G.L.J. (1975): Randomness and mathematical proof. In: Scientific American, May, pp 47-52.

Chen, P.P.-S. (1976): The entity-relationship model - Toward a unified view of data. In: ACM Trans. database systems, v.1 n.1, pp 9-36.

Chen, P.P.-S. (1982): Applications of the Entity-Relationship Model. In Yao, S.B. et al. (eds.): Data base design techniques. Vol. I. Berlin (Springer), pp 87-113.

Chen, Y.S., Ives, B., Leleux, D. (1997): Information Systems Knowledge Repository: A Structure for Electronic Scholarship. 3rd AIS Conference. http://hsb.baylor.edu/ramsower/ais.ac.97/papers/yschen.htm, (4/27/98).

Chicchetti, D. (1991): The reliability of peer review for manuscript and grant submissions. A cross-disciplinary investigation. In: Behavioral and brain sciences, vol. 14, pp 119-135.

Chmielewicz, K. (1994): Forschungskonzeptionen der Wirtschaftswissenschaft. 3rd ed. Stuttgart (Schäffer-Poeschel).

Cilliers, P. (1998: Complexity and postmodernism – Understanding complex systems. London (Routledge),

Cinkosky, J.; Fickett, J.W.; Gilna, P.; Burks, C. (1991): Electronic data publishing and GenBank. Science 252: 1273-1277.

Conklin, J. (1987): Hypertext: An introduction and survey. In: IEEE Computer, vol. 20 no. 9, pp 17-40.

Connolly, J.M.; Keutner, T. (1988): Hermeneutics versus science? Notre Dame (Univ. Notre Dame Press).

Cook, T.D.; Campbell, D.T. (1979): Quasi-experimentation: Design and analysis issues for field settings. Rand McNally (Chicago).

Cummings, A.M.; Witte, M.L.; Bowen, W.G.; Lazarus, L.O.; Ekman, R.H. (1992): University libraries and scholarly communication. (Association of Research Libraries).

CYCORP (1999): Background on the Cyc Technology. WWW document. http://www.cyc.com/tech.html (6/18/1999).

Daily, B.; Whatley, A.; Ash, S.R.; Steiner, R.L.(1996): The Effect of a Group Decision Support System on Culturally Diverse and Culturally Homongenous Group Decision Making. Information & Management, pp 281-289.

Dancy, J. (1985): An introduction to contemporary epistemology. Oxford (Blackwell).

DARPA (1991): Proceedings of the 3rd message understanding conference (MUC-3). San Diego (Morgan Kaufmann).

DARPA (1992): Proceedings of the 4th message understanding conference (MUC-4). MacLean, Va (Morgan Kaufmann).

Davenport, T.H.; Prusak, L. (1998): Working knowledge. Boston (Harvard Business School Press), pp 144-161.

Davies, J.T. (1965): The scientific approach. London (Academic Press).

Davis, F.D. (1989): Perceived usefulness, perceived ease of use, and user acceptance of information technology. In: MISQ, Sept. 1989, pp 319-339.

Daviss, B. (1999): Paper goes electric. In: New Scientist. Vol. 162 No. 2186 (5/15/1999), pp 36-39.

Dennis, A.R.; Hilmer, K.M.; Taylor, N.J. (1998): Information Exchange and Use in GSS and Verbal Group Decision Making: Effects of Minority Influence. Journal of Management Information Systems (14:3), pp 61-88.

Dennis, A.R.(1996): Information Exchange and Use in Group Decision Making: You Can Lead a Group to Information But You Can't Make It Think. MIS Quarterly, December 1996, pp 433-457.

DeRosnay, J. (1979): The macroscope. New York (Harper). Full text available at: http://pespmc1.vub.ac .be/MACRBOOK.html.

Director, S.; King, J. (eds.) (2001): President's information revolution commission report. University of Michigan, April 2001. http://www.umich.edu/pres/inforev/index.html (5/10/2001).

Descartes, R. (1927): Selections. Chicago (Scribner's sons).

Dolin, R.; Agrawal, D.; El Abbadi, A.; Pearlman, J. (1998): Using automated classification for summarizing and selecting heterogenous information sources. In: DLib, Jan. 1998, http://www.dlib.org/dlib /january98/dolin/01dolin.html.

Dubin, R. (1978): Theory building. New York (Free Press).

Dunnette, M.D. (1978): Fads, fashions and folderol in psychology. In: Dubin, R. (1978), pp 268-287.

Eldredge, N.; Cracraft, J. (1980): Phylogenetic patterns and the evolutionary process. New York (Columbia Univ. Press).

Ellison, G. (2000): Evolving standards for academic publishing. A q-r theory. National Bureau of Economic Research Working Paper No. 7805.

El Sawy, O.; Bowles, G. (1997): Redesigning the Customer Support Process for the Electronic Economy: Insights From Storage Dimensions. In: MISQ, vol. 21, no. 4, pp 457-483.

Eisenstein, E.L. (1983): The printing revolution in early modern Europe. Cambridge (Cambridge Univ. Press).

Emory, C.W. (1985): Business research methods. 3rd ed. Homewood (Irwin).

Engelbart, D.C. (1995): Toward augmenting the human intellect and boosting our collective IQ. Comm. ACM (38:8), pp 30-33.

Endres, F.F.; Wearden, S.T. (1990): Demographic portrait of journal reviewers. In: Educator, vol. 45, pp 45-52.

Farquhar, A.; Fikes, R.; Rice, J. (1996): The Ontolingua Server: A tool for collaborative ontology construction. Stanford Knowledge Systems Laboratory Report KSL-96-26. ftp://ftp-ksl.stanford.edu/pub/KSL_Reports/KSL-96-26.ps

Farquhar, A.; Fikes, R.; Pratt, W.; Rice, J. (1995): Collaborative ontology construction for ontology creation. Stanford Knowledge Systems Laboratory Report KSL-95-63. ftp://ftp.ksl.stanford.edu/pub/KSL_Reports/KSL-95-63.ps

Fay, B. (1987): Critical social science. Ithaca (Cornell University Press).

Feigl, H.; Sciven, M.; Maxwell, G. (eds.) (1958): Minnesota studies in the philosophy of science, vol. II, Concepts, theories and the mind-body problem. Minneapolis (Univ. Minnesota Press).

Finholt, T.A. and Olson, G.M. (1997): From laboratories to collaboratories - a new organizational form for scientific collaboration. Psychological Science, (8:1), pp 28-36.

Fisher, J.H. (1997): Comparing Electronic Journals to Print Journals: Are There Savings? Proceedings of the conference on Scholarly Communication and Technologies. http://arl.cni.org/scomm/scat /fisher.html (8/20/1998).

Fishwick, F.; Edwards, L.; Blagden, J. (1998): Economic implications of different models of publishing scholarly electronic journals for professional societies and other small or specialist publishers. Report Electronic Libraries Programme. http://www.ukoln .ac.uk/services/elib/papers/tavistock/scholarly-journals/cranelib.html (8/20/1998).

Fowler, M.; Scott, K. (1997): UML Distilled. Reading, MA (Addison-Wesley).
Franklin, B.J.; Osborn, H.W. (eds.) (1971): Research methods: Issues and insights. Belmont, CA (Wadsworth).
Freed, N.; Moore, K. (Nov. 1997): MIME Parameter Value and Encoded Word Extensions: Character Sets, Languages, and Continuations. Proposed Standard. The Internet Society RFC 2231, http://www.cis.ohio-state.edu/htbin/rfc/rfc2231.html (2/15/1999).
Frijhoff, W. (1996):Patterns. In: De Ridder-Symoens (H.) (ed.): A history of the University in Europe, Vol. II: Universities in early modern Europe (1500-1800). Cambridge (Cambridge University Press), pp 43-110.
Gadamer, H.G. (1976): Truth and method. New York (Seabury).
Gaines, B.R. (1996): Dimensions of electronic journals. In: Harrison & Stephen, 1996, pp 315-334.
Garfield, E. (1989): Citation behavior – An aid or hindrance to information retrieval. In: Current contents, 1 May, pp 3-8.(unread)
Garvey, W.D., & Griffith, B.C. (1972). Communication and information processing within scientific disciplines: Empirical findings for psychology. Information Storage and Retrieval, 8, 123–126.
Gefen, G.; Straub, D.W. (1997): Gender differences in the perception and use of e-mail: An extension to the technology acceptance model. In: MISQ, vol. 21, no. 4.
Gibbs, J.P. (1972): Sociological theory construction. Hinsdale, IL (Dryden).
Ginsparg, P. (1994): Physics preprint e-archive: First Steps Towards Electronic Research Communication. Computers in Physics, Vol.8, No.4, Jul/Aug 1994, pp 390-396.
Ginsparg, P. (1996): Winners and losers in the global research village. Presented at UNESCO conference, Electronic publishing in science, February, 1996, http://xxx.lanl.gov/blurb/pg96unesco.html (4/26/1998).
Glogoff, S. (1988): Reviewing the gatekeepers: A survey of referees of library journals. In: Journal of the American Society for Information Science, vol. 39, pp 400-407. (Unread)
Goldstein, M.; Goldstein, F. (1978): How we know. New York (Plenum).
Gordon, M.D. (1977): Evaluating the evaluators. In: New Scientist, vol. 73, pp 342-343. (Unread)
Grant, R.M. (1996): Toward a knowledge-based theory of the firm. In: Strategic Mangement Journal. Vol. 17 (Winter special issue), pp 109-122.
Greer, S. (1971): On the selection of problems. In: Franklin, B.J.; Osborn, H.W. (eds.): Research methods: Issues and insights. Belmont, CA (Wadsworth), pp 96-100, see p.96.
Grochla, E. (1974): Integrierte Gesamtmodelle der Datenverarbeitung. München.
Gruber, T.R. (1995): Towards principles for the design of ontologies for knowledge sharing. Int. Journal of Human-Computer Studies, 43(5/6), pp 907-928.
Guedon, J.-C. (1996): The seminar, the encyclopedia, and the eco-museum as possible future forms of electronic publishing. In: Peek, R.P. Newby, G.B. (eds.): Scholarly publishing - The electronic frontier. Cambridge (MIT Press), pp 71-89.
Guida, G.; Tasso, C. (1994): Design and development of knowledge-based systems. Chichester (Wiley).
Guha, R.V. (1998): Meta Content Framework : A Whitepaper. Draft. http://www.xspace.net/hotsauce /wp.html (8/29/1998).
Gutenberg, E. (1965): Grundlagen der Betriebswirtschaftslehre. Erster Band: Die Produktion. 10th ed., Berlin (Springer).
Guttenplan, S. (1997): The languages of logic. 2nd ed. Oxford (Blackwell).

Habermas, J. (1981): The theory of communicative action. Boston (Beacon).
Habermas, J. (1968): Knowledge and human interests. Boston (Beacon).
Halpern, J.Y. (1998): A computing research repository. In: D-Lib Magazine (Nov. 1998). http://www.dlib.org/dlib/november98/11halpern.html.
Hammer, M. and Champy, J. (1994): Reengineering the Corporation. Harper Business.
Hamilton, D.P. (1990): Publishing by – and for? – the numbers? In: Science 250, 1331-32.
Harnad, S. (1992): Interactive Publication: Extending American Physical Society's Discipline-Specific Model for Electronic Publishing. Serials Review, Special Issue on Economics Models for Electronic Publishing, pp 58 - 61.
Harnad, S. (1995): Publicly retrievable FTP archives for esoteric science and scholarship: a subversive proposal. In: Okerson, A.; O'Donnel, J. (eds.): Scholarly journals at the crossroads: A subversive proposal for electronic publishing. Washington, (Ass. of research libraries).
Harnad, S. (ed.) (1982): Peer commentary on peer review: A case study in scientific quality control. Reprinted from The Behavioral and Brain Sciences, vol.5, no.2. Cambridge UK, (Cambridge Univ. Press).
Hars, A.(1998): Natural language-based data modeling: Improving validation and integration. In: Journal of Database Management, vol. 9, no. 2 (Spring 1998), pp 17-25.
Hars, A. (1994): Referenzdatenmodelle - Grundlagen effizienter Datenmodellierung. Wiesbaden (Gabler), 1994.
Hars, A.; Malhotra, A. (1997): Emerging web-based knowledge bases: Categories, trends and implications for the IS research community. Unpublished manuscript.
Hars, A.; El Sawy, O.; Gosain, S.; Hirt, S.; Im, I.; Kang, D.; Lee, Z.; Raven, A.: Reengineering information systems research for the electronic economy. Journal of Electronic Commerce, vol. 10 (2000), no. 2, pp 67-83.
Hars, A.; Im, I.(1998): Cumulative knowledge in information systems research: Reality vs. opportunity. USC Information Systems Working paper. (Submitted to ICIS 1998)
Harrison, T.M.; Stephen, T. (eds.) (1996): Computer networking and scholarly communication in the Twenty-First Century. New York (State University of New York Press).
Hart, M.H. (1992): The 100: A ranking of the most influential persons in history. New York (Citadel Press).
Harter, S.P.; Ford, C.E. (2000): Web-based analyses of E-journal impact: Approaches, problems and issues. In: JASIS, vol. 51, no. 13, pp 1159-1176.
Harter, S.P.; Park, T.K. (2000): Impact of prior electronic publication on manuscript consideration policies of scholarly journals. In: JASIS, vol. 51, no. 10, pp 940-948.
Hempel, C.G. (1958): The theoretician's dilemma. In: Feigl, Scriven, Maxwell (1958), pp 37-98.
Heinemann Pieper (1988): Comments on "Research paradigms in social work: From stalemate to creative synthesis". In social service review, vol. 62, no. 3, pp 535-536.
Heylighen, F.; Bollen, J. (1995): Development and Publication of Systems Knowledge on the Internet: The Principia Cybernetica Web. Manuscript. ftp://ftp.vub.ac.be/pub /projects/Principia_Cybernetica/ Texts_General/PCP_Web.txt (12/27/1998).
Hildredth, C.R. (1996): Preserving what we really want to access: The message, not the medium: Challenges and opportunities in the digital age. In Helal, A.H.; Battin, P. (eds.): Electronic documents and information: from preservation to access. Essen (Essen Univ. Library), pp 78-95.
Hilgartner, S. (1995): Biomolecular databases – New communications regimes for biology? In: Science communication, vol. 17, no. 2 (December), pp 240-263.

Hitchcock, S.; Carr, L.; Hall, W. (1997): Web journals publishing: a UK perspective. http://journals.ecs.soton.ac.uk/uksg.htm (10/12/98).
van Hoesel, C.P.M; Wagelmans, A.P.M. (1996): An O(T^3) algorithm for the economic lot-sizing problem with constant capacities. In: Management Science, vol. 43, no. 1 (January), pp 142-150.
Horrobin, D.F. (1982): Peer review: A philosophically faulty concept which is proving disastrous to science. Commentary. In: Harnad (1982), pp 33-34.
Horowitz, I.L. (1986): Communicating ideas. The crisis of publishing in a post-industrial society. Oxford (Oxford Univ. Press).
Hume, D. (1748): Sceptical doubts concerning the operations of the human understanding. Reprinted in: Bierman & Gould (1981), pp 176-194.
Hurd, J.M. (2000): The transformation of scientific communication: A model for 2020. In: JASIS, vol. 51, no. 14, pp 1279-1283.
Isakowitz, T.; Stohr, E.A.; Balasubramian, P. (1995): RMM: A methodology for structured hypermedia design. In: Comm. ACM. Vol. 38. no. 8, pp 34-44.
Jackson, D. (1981): The story of writing. New York (Taplinger).
Ives, B. (1994): MISQ Central: Creating a new intellectual infrastructure. Editors Comments, MISQ Vol. 18, No. 3, pp xxxv-xxxix.
Ives, B.; Zmud, B. (1994): ISWorldNet: A scholarly infrastructure for information systems. Editor's comments. MISQ, Vol. 18, no. 4, pp liv-lvi.
Jacobson, J. et al. (1998): The last book. In IBM Systems Journal. Vol. 36, no. 3, pp 457-463. .
James, W. (1911): Some problems in philosophy: Percept and concept and their practical uses. Reprinted in: Bierman & Gould (1981), pp 185-194.
Jarvenpaa, S.L. (1990): Panning for gold in information systems research: 'Second-hand' data. In: Nissen et al. (1990), pp 17-32.
Jauch, L.R., Wall, J.L. (1989): What they do when they get your manuscript: A survey of Academy of Management Reviewer practices. In: Academy of Management Journal, vol. 32, pp 157-173.
Kalakota, R.; Whinston, A.B. (1996): Frontiers of electronic commerce. Reading, MA (Addison Wesley).
Kant, I. (1787): Critique of pure reason. Translated by Kemp Smith, New York (St. Martin's Press).
Kelly, G.G.; Bostrom, R.P. (1998): A Facilitator's General Model for Managing Socioemotional Issues in Group Support Systems Meeting Environments. JMIS (14:3), pp 23-44.
Kling, R.; Covi, L. (1995): Electronic journals and legitimate media in the systems of scholarly communication. In: Information Society, vol. 11, no. 4, pp 261–271.
Kling, R.; Lamb, R. (1996): Analyzing alternative visions of electronic publishing in digital libraries. In: Peek & Newby (1996), pp 17-54.
Kogut, B.; Zander, U. (1992): Knowledge of the firm: combinative capabilities, and the replication of technology. In: Organization Science, vol. 3, no. 3, pp 393-397.
Knorr-Cretina, K.D. (1981): The manufacture of knowledge. Oxford (Pergamon).
Kristen, G. (1994): Object-orientation: The KISS method. Wokingham.
Kuhn, T.S. (1996): The structure of scientific revolutions. 3rd ed. Chicago (Univ. of Chicago Press).
Kurzweil, Raymond (1990): The age of intelligent machines. Cambridge (MIT Press).

Labov, W. (1973): The boundaries of words and their meanings. In: Bailey, C.-J.N.; Shuy, R.W. (eds.): New ways of analysing variation in English. Washington (Georgetown University Press), pp 340-373.

LaFollette, M.C. (1992): Stealing into print. Berkely (Univ. of California Press).

Lam, S.S.K (1997): The Effects of Group Decision Support Systems and Task Structures on Group Communication and Decision Quality. Journal of Management Information Systems (13:4), pp 193-215.

Lawrence, S.; Giles, C.L. (1998): Searching the world wide web. Science, vol. 280, 3/4/1998, pp 98-100.

Laudan, L. (1996): Beyond positivism and relativism: Theory, method and evidence. Boulder (Westview).

Lederberg, J.; Uncapher, K. (1989): Towards a national collaboratory: Report of an invitational workshop at the Rockefeller University. Washington, D.C.: NSF, Directorate for Computer and Information Science and Engineering.

Lenat, D.B. (1995): CYC: A large scale investment in knowledge infrastructure. In CACM vol. 38, no. 11, pp 32-38.

Lerner, R.G. (1980): Access to data in the primary literature. In Rossmassler, S.A.; Watson, D.G. (eds.): Data handling for science and technology. Amsterdam (North Holland), pp 75-82.

Lewin, K. (1945): The research center for group dynamics at MIT. In: Sociometry, vol. 8, pp 126-135.

Lewis, D.D.; Jones, K.S. (1996): Natural language processing for information retrieval. In: Comm. ACM, vol. 39, no. 1, pp 92-101.

Lim, K.H., and Benbasat, I. (1997): A Framework for Addressing Group Judgment Biases with Group Technology. Journal of Management Information Systems (13:3), pp 7-24.

Lim, K.H.; Ward, L.M.; Benbasat, I. (1998): An Empirical Study of Computer System Learning: Comparison of Co-discovery and Self-discovery Methods. Information Systems Research, (8:3), pp 254-272.

Lifschitz, V. (ed) (1990): Formalizing common sense. Papers by John McCarthy. Norwood (Ablex).

Little, D. (1998): Microfoundations, method and causation: On the philosophy of the social sciences. New Brunswick (Transaction Publ.).

Lock, S. (1985): Letter to P.B.S. Fowler 12/4/1984. In: British Medical Journal, vol. 290, p 1560.

Logoze, C.; Fielding, D. (1998): Defining collections in distributed digital libraries. In D-Lib Magazine, Nov. 1998. www.dlib.org/dlib/november98/lagoze/11lagoze.html.

Mahoney, M.J. (1985): Open exchange and epistemic progress. In: American Psychologist, vol. 40, pp 29-39.

Mann, C.E. (2001): Electronic paper turns the page. In: Technology Review, vol.104, no.2, pp 42-48.

Marcuse, H. (1968): Reason and revolution. London (Routledge). (Unread)

Martin, J. (1990): Information Engineering. Englewood Cliffs (Prentice Hall).

Mazur, A. (1971): The littlest science. In Franklin, B.J; Osborne, H.W. (eds.): Research methods: Issues and insights. Belmont (Wadsworth), pp 14-20.

McCarthy, J. (1990): An example for natural language understanding and the AI problems it raises. In: Lifschitz (1990), pp 70-76.

McCarthy, J., Hayes, P. (1969): Some philosophical problems from the standpoint of artificial intelligence. In Meltzer, B.; Michie, D. (eds.): Machine Intelligence 4. Edingburgh (Edingburgh Univ. Press) pp 463-502, reprinted in Lifschitz (1990) pp 21-63.

McDavid, D.W. (1996): Business language analysis for object-oriented information systems. In: IBM Systems Journal, vol. 35, no. 2, pp 128-150.

McKnight, C.; Dillon, A.; Shackel, B. (1996): The electronic journal and its implications for the electronic library. In: Harrison, Stephen (1996), pp 351-368.

McNutt, R.A. et al. (1990): The effects of blinding on the quality of peer review. In: Journal of the American Medical Association, vol. 263, pp 1371-1374.

Menard, H.W. (1971): Science: Growth and change. Cambridge (Harvard).

Miller, D.C. (1991): Handbook of Research Design and Social Measurement. 5th ed. Newbury Park, CA (Sage).

Miller, D. W. (1997): On methodological proposals. In: Keuth, H. (ed.): Karl Popper, Logik der Forschung. Berlin (Akademie), pp 67-81.

Miller, R.W. (1991): Fact and method in the social sciences. In: Boyd et al. 1991, pp 743-162.

Miller, P.McC.; Wilson, M.J. (1983): A dictionary of social science methods. Chichester (Wiley).

Mintzer, F.; Braudaway, G.W.; Bell, A.E. (1998): Opportunities for watermarking standards. In: Comm. ACM, vol. 41, no. 7, pp 57-64.

Mitev, N. (2000): Toward social constructivist understandings of is success and failure: Introducing a new computerized reservation system. In: Orlikowski, W. et al. (eds.): Proc. 21st ICIS, Brisbane, pp 84-93.

Moore, G.E. (1903): Principia ethica. Cambridge (Cambridge Univ. Press), p 10.

Münz, S. (1996): Hypertext. World-wide web document. http://www.netsystems.ch/ support/hypertext /htxt.htm (9/11/1998).

Murray, G.D. (1988): The task of a statistical referee. In: British Journal of Surgery, vol. 75, pp 664-667.

National Research Council (1993): National collaboratories - Applying information technology for scientific research. Washington (National Academy Press).

Nagel, E. (1961): The structure of science. New York (Harcourt, Brace & York).

Nelson, R. (1991): Why do firms differ, and how does it matter? In: Strategic Management Journal. Vol. 12, Winter special issue, pp 61-74.

Nelson, R. E.; Mathews, K. M. (1991): Cause maps and social network analysis in organizational diagnosis. In: Journal of Applied Behavioral Science, vol. 27, no. 3.

Nelson, T. (1981): Literary machines. Swathmore (self-published).

Nijssen, G.M.; Halpin, T.A. (1989): Conceptual schema and relational database design. A fact oriented approach. New York.

Nissen, H.E.; Klein, H.K.; Hirschheim, R. (1990) (eds.): The information systems research arena of the 90s. Proceedings of the IFIP TC8 WG2 Working Conference, Copenhagen.

Norman, R.J. (1996): Object-oriented systems analysis and design. Upper Saddle River, NJ (Prentice Hall).

O'Connor, M. (1986): How to copyedit scientific books and journals. Philadelphia (ISI Press).

Ocker, R.; Hiltz, S.R.; Turoff, M.; Fjermestad, J. (1995): The Effect of Distributed Group Support and Process Structuring on Software Requirements Development Teams: Re-

sults on Creativity and Quality. Journal of Management Information Systems (12:3), pp 127-153.
Odlyzko, A. (1997): The Economics of Electronic Journals. Proceedings of the conference on Scholarly Communication and Technologies. http://arl.cni.org/scomm/scat/odlyzko.html (8/20/1998).
Okerson, A. (1996): University libraries and scholarly communication. In: Peek & Newby (1996), pp 181-199.
Okerson, A. (1995) (ed.): Scholarly publishing on the electronic networks: Filling the pipeline and paying the piper. Proceedings of the 4th symposium, Nov. 1994. Washington (Assoc. of Research Libraries).
Okerson, A. (1993) (ed.): Visions and opportunities in electronic publishing. Proceedings of the 2nd symposium Dec. 1992. Washington (Assoc. of Research Libraries).
Olle, T.W. et al. (1991): Information systems methodologies – A framework for understanding. 2nd ed. Wokingham.
Olson, G.M.; Olson, J.S. (1997): Research on Computer Supported Cooperative Work. In: Helander, M., Landauer, T.K.; Prabhu, P. (eds).: Handbook of Human-computer Interaction. 2nd ed., Amsterdam (Elsevier), chapter 59, pp 1433-1456.
Olson, G.M.; Olson, J.S.(1996): Groupware and Authoring. Academic Press.
Pasa, M.; Shugan, S.M. (1996): The value of marketing expertise. In: Management science, vol. 43, no. 3 (March), pp 371-388.
Peek, R.P.; Newby, G.B. (1996): Scholarly publishing. Cambridge (MIT Press).
Peile, C. (1994): The creative paradigm. Avebury Series in Philosophy. Aldershot, UK (Avebury).
Peters, D.P.; Ceci, S.J. (1982): Peer review practices of psychological journals: The fate of published articles, submitted again. In Harnad (1982), pp 3-11.
Popper, K.R. (1968): The logic of scientific discovery. 3rd ed. London (Hutchinson).
Pratt, V. (1994): CYC Report. http://boole.stanford.edu/pub/cyc.report (7/30/1998).
Price, D.J. (1986): Little Science, Big Science and Beyond. New York (Columbia Univ. Press).
REFA (ed.) (1978): Methodenlehre des Arbeitsstudiums. Vol. 2 Datenermittlung. 6th ed. München.
Reichenbach, H. (1938): Experience and prediction. Chicago (Univ. of Chicago Press). PHILOSOPHY BD 163 R4
Remenyi, D.;Williams, B.; Money, A.; Swartz, E. (1998): Doing research in business and management. An introduction to process and method. London (Sage).
Riecken, H.W. (1989): Scholarly publications: Are there viable options? Draft for the Research Library Committee of the Council on Library Resources. October, 1989.
Riloff, Eleanor (1994): Information extraction as a basis for portable text classification systems. Dissertation, Univ. of Massachussets, Amherst.
Rosch, E. (1973): On the internal structure of perceptual and semantic categories. In: Moore, T.E. (ed.): Congnitive development and the acquisition of language. New York (Academic Press), pp 111-144.
Rosemann, M. (1996): Komplexitätsmanagement in Prozessmodellen. Wiesbaden (Gabler), 1996.
Rossmassler, S.A. (1980): Presentation of data in the primary literature. In Rossmassler, S.A.; Watson, D.G. (eds.): Data handling for science and technology. Amsterdam (North Holland), pp 65-69.

Ross-Flanigan, N. (1998): The virtues (and vices) of virtual colleges. In: MIT Technology Review, March-April, pp 53-59. http://www.techreview.com/articles/ma98/ross-flanigan.html.

Rumbaugh, J.; Blaha, M.; Premerlan, W.; Eddy, F.; Lorensen, W. (1991): Object-oriented modeling and design. Englewood Cliffs, NJ (Prentice Hall).

Russel, B. (1918): Mysticism and logic and other essays. London (Longmann, Green). Reprinted in: Slater, J.G. (ed.) (1992): Bertrand Russel: Logical and philosophical papers. 1903-1913. Vol. 6. (Routledge), pp 147-161.

Schank, R.C.; Abelson, R.P. (1977): Script, plans, goals and understanding. New York (Wiley).

Schauenberg, B. (1998): Gegenstand und Methoden der Betriebswirtschaftslehre. In Dellman, K. et al. (eds.): Vahlens Kompendium der Betriebswirtschaftslehre. Vol.1. München (Vahlen), pp 1-56.

Schlick, M. (1974): General theory of knowledge. New York (Springer).

Scheer, A.-W. (1998a): ARIS - Business process frameworks. 2nd. edition, Berlin (Springer).

Scheer, A.-W. (1998b): ARIS – Business process modeling. 2nd ed. Berlin (Springer).

Schmidt, G. (1996): Informationsmanagement. Berlin (Springer).

Schwab, D.P. (1980): Construct validity in organizational behavior. In: Staw, B.M.; Cummings, L.L. (eds.): Research in organizational behavior. Vol. 2, no. 2, pp 3-43. Greenwich, CT (JAI Press).

Schwartz, C. (1998): Web search engines. In: Journal of the American Society for Information Science, vol. 49, no. 11, pp 973-982.

Schwemmer, O. (1981): Die Vernunft der Wissenschaft – Kritische Anmerkungen zu einem unvermeidlichen Anspruch. In: Janich, P. (ed.): Wisenschaftstheorie und Wissenschaftsforschung. München (Beck), pp 52-88.

Scriven, M. (1958): Definitions, explanations and theories. In: Feigl, Scriven, Maxwell (1958), pp 99-195.

Shannon, C.E.; Weaver, W. (1949): The mathematical theory of communication. Urbana, (Univ. of Illinois Press).

Shepherd, M.M.; Briggs, R.O.; Reinig, B.A.; Yen, J.; Nunamaker, J.F. (1995): Invoking Social Comarison to Improve Electronic Brainstorming: Beyond Anonymity. Journal of Management Information Systems (12:3), pp 155-170.

Shipman, F.M. et al. (2000): Guided paths through web-based collections: Design, experiences and adaptations. In: JASIS, vol. 51, no. 3, pp.260-272.

Simon, R.J.; Fyfe, J.J. (eds.) (1994): Editors as gatekeepers: Getting published in the social sciences. Boston (Rowman and Littlefield).

Simon, R.J.; Bakanic, V.; McPhail, C. (1986): Who complains to the journal editor and what happens. In: Sociological inquiry, vol. 56, pp 259-271.

Sirapayan, N. et al. (2000): Seach Engines. PC Magazine Review. Nov, 15, 2000. http://www.zdnet.com/pcmag/stories/reviews/0,6755,2652815,00.html (12/15/2000).

Spring, M.B. (1991): Electronic printing and publishing. New York (Marcel Dekker).

Sowa, J.F. (1995): Top-level ontological categories. In: Int. J. human computer studies, vol. 43, pp 669-685.

Sowa, J.F.; Zachman, J.A. (1992): Extending and formalizing the framework for information systems architecture. In: IBM Systems Journal, vol. 31, pp 590-619.

Star, S.L.; Ruhleder, K. (1996): Steps toward an ecology of infrastructure: design and access for large information spaces. In: Information Systems Research, vol. 7, no. 1, pp 111-134.

Stegmüller, W. (1976): The structure and dynamics of theories. New York (Springer).

Sternberg, R.J. (1985): Tacit agreements between authors and editors. In: Behavioral and brain sciences, vol. 8, pp 746-747.

Stein, E. W. (1992): A method to identify candidates for knowledge acquisition. In: Journal of Management Information Systems, fall 1992.

Steinberg, S.H. (1959): 500 years of printing. London (Faber & Faber).

Story, G.A.; O'Gorman, L.; Fox, D.; Levy Schaper, L.; Jagadish, H.V. (1992): The RightPages image-based electronic library for alerting and browsing. IEEE Computer, vol. 25, no. 9., pp 17-26.

Strike, K.; Posner, G. (1983): Types of synthesis and their criteria. In: Ward & Reed (1983), pp 343-362.

Sullivan, D. (2001): Search Engines Sizes. Search Engine Watch Report, 4/6/2001, http://www.searchenginewatch.com/reports/sizes.html (5/1/2001).

Sullivan, D. (1998): The Altavista Controversy. http://www.searchenginewatch.com/facts/avsize.html (8/28/1998).

Sutherland, J.W. (1975): Systems: Analysis, administration, and architecture. New York (Van Nostrand).

Sztompka, P. (1990): Conceptual frameworks in comparative inquiry: Divergent or convergent? In: Albrow, M.; King, E. (1990): Globalization, knowledge and society. London (Sage), pp 47-58.

Swartz, M.E.; Krull I.S. (1997): Analytical method development and validation. New York (Dekker).

Taylor, J.R. (1995): Linguistic categorization - Prototypes in linguistic theory. 2nd ed., Oxford (Oxford University Press).

Tenopir, C.; King, D. (2000): Towards electronic journals: Realities for scientists, librarians and publishers. Washington (Special Libraries Association).

Thorndike, R.L; Hagen, E. (1969): Measurement and evaluation in pschology and education, 3rd ed. New York (Wiley).

Treloar, A. (1998): Libraries' new role in electronic scholarly publishing. In: Comm. ACM, vol. 41, no. 4, pp 88-89.

Turner, J.H. (1974): The structure of sociological theory. Homewood, IL (Dorsey).

Turner, J.H. (ed), (1986): Theory building in sociology: Assessing theoretical cumulation. Newbury Park, CA (Sage).

Uschold, M. (1996): Building ontologies: Towards a unified methodology. Manuscript, Artificial Intelligence Applications Institute, University of Edinburgh.

Uschold, M.; Grüninger, M. (1996): Ontologies: Principles, methods and applications. Technical report AIAI-TR-191 of the Artificial Intelligence Applications Institute, University of Edinburgh.

Vandenbosch, B.; Ginzberg, M.J. (1997): Lotus Notes and Collaboration: Plus ca Change. Journal of Management Information Systems (13:3), pp 65-61.

Van de Sompel, H.; Lagoze, C. (2001): The Open Archives Initiative protocol for metadata harvesting. Version 2001-01-21. http://www.openarchives.org/OAI/ openarchivesprotocol.htm (2/20/2001).

Van Doren, C. (1991): A history of knowledge. New York (Ballantine).

Varian, H. (1997): The future of electronic journals. Proceedings of the conference on Scholarly Communication and Technologies. http://arl.cni.org/scomm/scat/varian.html (8/20/1998).
Wagner, Chr. (1989). View integration in database design. PhD thesis, University of British Columbia, Vancouver.
Walls, J.G., Widmeyer, G.R., El Sawy, O. (1992): Building an information systems design theory for vigilant EIS. In: Information Systems Research vol. 3, no.1, pp 36-59.
Watson, R. (1994): Creating and sustaining a global community of scholars. In: MISQ, Vol. 18, No 3, 228-231, hwww.misq.org/archivist/vol/no18/issue3/vol18n3art1watson.html.
Weibel, S.; Hakala, J. (1998): DC-5: The Helsinki Metadata Workshop. In: D-Lib Magazine, February, 1998. http://www.dlib.org/dlib/february98/02weibel.html.
Weick, K.E. (1989): Theory construction as disciplined imagination. In: Academy of Management Review, vol. 14, no.4, pp 516-531.
Whetten, D.A. (1989): What constitutes a theoretical contribution? In: Academy of Management Review, vol. 14, no. 4, pp 490-495.
Weller, A.C. (2000): Editorial Peer Review for Electronic Journals: Current Issues and Emerging Models. In: JASIS vol. 51, no. 14, pp 1328-1333.
Wilson, J.D. (1978): Peer review and publication: Presidential address for the 70st annual meeting of the American Society for Clinical Investigation, San Francisco, California, April 30 1978. In: Journal of Clinical Investigation, vol. 67, pp 1697-1701.
Wittgenstein, Ludwig (1978): Philosophical investigations. Translated by Anscombe, G.E.M. Oxford (Basil Blackwell).
Wulf, W. (1989): The national collaboratory – A whitepaper. In: Lederberg, Uncapher (1989), Appendix A.
Yankauer, A. (1991): Who are the peer reviewers and how much do they review? In: Journal of the American Medical Association, vol. 263, pp 1679-1701.
Ytterstad, P.; Akselsen, S.; Svendsen, G.; Watson, R.T. (1996): Teledemocracy: Using Information Technology to Enhance Political Work. MISQ Discovery, September 1996, http://www.misq.org/discovery/articles96/article1/.
Zachmann, J.A. (1987): A framework for information systems architecture. In: IBM systems journal, vol. 26, no. 3, pp 276-292.
Zigurs, I.; Kozar, K.A. (1994): An Exploratory Study of Roles in Computer-Supported Groups. MIS Quarterly, September 1994, pp 277-297.
Ziman, J. (1968): Public knowledge: The social dimension of science. Cambridge (Cambridge Univ. Press).

Druck: Strauss Offsetdruck, Mörlenbach
Verarbeitung: Schäffer, Grünstadt